Essentials of Nucleic Acid Analysis
A Robust Approach

Essentials of Nucleic Acid Analysis
A Robust Approach

Edited by

Jacquie T. Keer and Lyndsey Birch
LGC, Teddington, Middlesex, UK

RSCPublishing

ISBN: 978-0-85404-367-5

A catalogue record for this book is available from the British Library

© LGC Limited 2008

All rights reserved

Apart from fair dealing for the purposes of research for non-commercial purposes or for private study, criticism or review, as permitted under the Copyright, Designs and Patents Act 1988 and the Copyright and Related Rights Regulations 2003, this publication may not be reproduced, stored or transmitted, in any form or by any means, without the prior permission in writing of The Royal Society of Chemistry, or the Copyright owner, or in the case of reproduction in accordance with the terms of licences issued by the Copyright Licensing Agency in the UK, or in accordance with the terms of the licences issued by the appropriate Reproduction Rights Organization outside the UK. Enquiries concerning reproduction outside the terms stated here should be sent to The Royal Society of Chemistry at the address printed on this page.

Published by The Royal Society of Chemistry,
Thomas Graham House, Science Park, Milton Road,
Cambridge CB4 0WF, UK

Registered Charity Number 207890

For further information see our website at www.rsc.org

Preface

The last two decades have seen an explosion in the use of DNA analysis, with key applications encompassing forensic science, pathogen identification, food authenticity and detection of GMOs, personalised medicine and medical diagnostics. Its broad utility has encouraged a rapid and sustained development of the technology, with a wide range of techniques and products being introduced each year as well as new technologies emerging from the research base.

Although many of the commercial offerings help the analyst, DNA analysis remains a complex multi-step process and achieving a valid result is by no means a trivial task. This book sets out to guide the analyst through the steps needed to obtain good quality results. The underlying principles for achieving this goal were formulated by LGC as the six principles for ensuring valid analytical measurement, which are detailed in the Introduction. How to apply these principles to DNA analysis is a core feature of the book. The authors of each Chapter are practitioners of the art of DNA analysis in areas where the quality of the result is critical, be it in forensic applications, food analysis or working at the highest international level, through LGC's role as the designated UK National Metrology Institute for chemical and biochemical measurements. Their advice is based on first-hand experience of making high-quality measurements, which takes the reader through the essential elements for making sound, valid DNA measurements, be they qualitative or quantitative. This updated volume covers topics such as qPCR and microarray analysis, but the underlying theme remains one of quality to ensure that the correct result is achieved first time.

The book is designed to serve as a key component in the DNA analyst's toolkit for designing, planning and carrying out high-quality DNA measurement.

Dr John Marriott
Government Chemist

Contents

Abbreviations xix

Acknowledgements xxiii

Chapter 1 Valid Analytical Molecular Biology: The Challenge
Jacquie T. Keer

    1.1 Introduction   1
    1.2 The Analytical Process   2
          1.2.1 Analytical Requirements   2
          1.2.2 Stages in the Analytical Process   3
    1.3 Principles Underpinning Reliable Measurement   4
          1.3.1 Understand the Experimental Requirements   5
          1.3.2 Use Methods and Equipment which are Fit for the Intended Purpose   5
          1.3.3 Staff Undertaking Analysis Should be Both Qualified and Competent to Undertake the Task   5
          1.3.4 Regular Independent Assessment of Laboratory Performance   5
          1.3.5 Analytical Consistency   6
          1.3.6 Quality Control and Quality Assurance Framework   6
    1.4 Challenges to Measurement Quality   6
          1.4.1 Low Concentration of Analyte Compared to Matrix   6
          1.4.2 Complex Matrices   7
          1.4.3 DNA Degradation   7
          1.4.4 Biological Contamination of the Sample   7
          1.4.5 Degradation of Matrix Components   7
          1.4.6 Limited Availability of the Sample   8
          1.4.7 Lack of Suitable Controls   8
    1.5 Focus on Data Quality   8
    Acknowledgements   9

Chapter 2 Quality in the Analytical Molecular Biology Laboratory
Sally L. Hopkins

2.1	Introduction		10
2.2	Management Systems		11
2.3	Internationally Recognised Assessed Standards		12
	2.3.1	ISO 9001:2000 Quality Management Systems – Requirements	14
	2.3.2	ISO/IEC 17025:2005 General Requirements for the Competence of Testing and Calibration Laboratories	15
	2.3.3	ISO 15189:2003 Medical Laboratories – Particular Requirements for Quality and Competence	16
	2.3.4	Principles of Good Laboratory Practice 1999 (GLP)	16
	2.3.5	Joint Code of Practice for Research	16
2.4	Selection and Implementation of a Formal Management System		17
	2.4.1	The Management System	18
		2.4.1.1 Quality Manual	19
		2.4.1.2 Quality Procedures (QPs)	20
		2.4.1.3 Standard Operating Procedures (SOPs)	20
		2.4.1.4 Locally Controlled Documentation	22
	2.4.2	Laboratory Environment	22
		2.4.2.1 Safety	22
		2.4.2.2 Spatial Separation	23
	2.4.3	Equipment	24
		2.4.3.1 Analytical Requirement	24
		2.4.3.2 'Ownership'	24
		2.4.3.3 Log Books and Maintenance	24
		2.4.3.4 Calibration	25
	2.4.4	Reagents	25
		2.4.4.1 Reagent Quality	25
		2.4.4.2 Storage Conditions	26
		2.4.4.3 Reagent Traceability	26
		2.4.4.4 Stability/Batch Comparability	26
	2.4.5	Analysts	26
		2.4.5.1 Culture and Competence	26
		2.4.5.2 Training and Development	26
	2.4.6	Methods	28
		2.4.6.1 Fitness for Purpose	28
		2.4.6.2 Documentation	28
		2.4.6.3 Metrological Traceability	28
		2.4.6.4 Independent Quality Assessment	28

		2.4.6.5	Method Validation	29
		2.4.6.6	Experimental Design	29
		2.4.6.7	Measurement Uncertainty	30
	2.4.7	Quality Control		30
		2.4.7.1	Reference Materials	31
		2.4.7.2	In-house Quality Control Materials	31
		2.4.7.3	Performance Control	32
		2.4.7.4	Contamination Control	33
	2.4.8	Samples		33
		2.4.8.1	Chain of Custody	33
		2.4.8.2	Sampling and Preparation	34
		2.4.8.3	Storage	34
	2.4.9	Recording and Reporting		35
		2.4.9.1	Electronic Data and Automated Analysis	35
		2.4.9.2	Reporting	36
	2.4.10	Archiving		36
		2.4.10.1	Electronic Data	37
2.5	Summary			37
Acknowledgements				38
References				38

Chapter 3 An Introduction to Method Validation
Sally L. Hopkins and Vicki Barwick

3.1	Introduction			40
	3.1.1	Why and When is Method Validation Necessary?		41
		3.1.1.1	Criticality of the Data	42
		3.1.1.2	Uniqueness of the Sample	42
		3.1.1.3	Robustness of the Technique	42
		3.1.1.4	Expected Level of Utilisation of the Technique	43
3.2	Planning the Validation Process			43
3.3	Method Performance Parameters			43
	3.3.1	Precision		44
		3.3.1.1	Repeatability	45
		3.3.1.2	Reproducibility	46
		3.3.1.3	Intermediate Precision	46
	3.3.2	Bias		46
	3.3.3	Recovery		47
	3.3.4	Accuracy		48
	3.3.5	Ruggedness (Robustness) Testing		49
	3.3.6	Selectivity		49
	3.3.7	Detection Limit (Sensitivity)		50

		3.3.8	Working Range and Linearity	51
		3.3.9	Measurement Uncertainty	52
	3.4	Validation in Practice		53
		3.4.1	Outline of the Procedure	53
			3.4.1.1 Define the Analytical Requirement	53
			3.4.1.2 Write a Draft Protocol	54
			3.4.1.3 Investigate the Robustness of the Technique, and Identify the Critical Parameters	54
			3.4.1.4 Identify Relevant Performance Parameters, and Determine the Order of Investigation	54
			3.4.1.5 Assess the Performance Characteristics Using Suitable 'Known' Materials (RMs, Standards, Spikes)	54
			3.4.1.6 Assess Whether the Data Show the Method is Fit for Purpose	55
			3.4.1.7 Define the Limitations of the Methodology	57
			3.4.1.8 Document the Final Protocol and Method Validation Results	57
	Acknowledgements			57
	References			57

Chapter 4 DNA Extraction
Ginny C. Saunders and Jennifer M. Rossi

	4.1	Introduction		59
		4.1.1	Concentration or Amount	60
		4.1.2	Purity	60
		4.1.3	Integrity	60
	4.2	Steps of the DNA Extraction Process		61
		4.2.1	Sample Preparation	61
		4.2.2	Cell or Membrane Lysis	61
		4.2.3	Protection and Stabilisation of Released DNA	61
		4.2.4	Separation of Nucleic Acids from Cell Debris or Sample Matrix	61
		4.2.5	Purification of DNA	61
		4.2.6	Concentration of DNA	62
	4.3	Choosing an Appropriate DNA Extraction Procedure		62
		4.3.1	The History of the Sample	62
		4.3.2	The Composition of the Sample	62
		4.3.3	Time and Resources Available	63
		4.3.4	Standardised Techniques	63

		4.3.5	Subsequent Analytical Procedures	63
		4.3.6	Potential Impact of Methodology	63
4.4	Validation Issues Arising at the Various Stages of DNA Extraction			66
	4.4.1	Sample Storage		66
		4.4.1.1	Incorrect Sample Storage Temperature	66
		4.4.1.2	Incorrect Sample Storage Environment	66
	4.4.2	Sample Preparation		66
		4.4.2.1	Homogeneity of Sample	67
		4.4.2.2	Surface Area to Lysis Forces Ratio	67
		4.4.2.3	Cell or Nucleic Acid Adherence to Matrix Material	67
		4.4.2.4	Contamination	67
	4.4.3	Cell and Membrane Lysis		68
		4.4.3.1	Inaccessibility of Cells to Lysis Forces	68
		4.4.3.2	Type and Amount of Detergent or Denaturant Used	68
		4.4.3.3	Concentration and Activity of Lytic Enzyme	68
		4.4.3.4	Concentration of EDTA in Extraction Buffer	69
		4.4.3.5	Concentration of Salt in Extraction Buffer	70
		4.4.3.6	Extraction Buffer pH	70
		4.4.3.7	Excessive Damage of the DNA Analyte	70
	4.4.4	Separation of Nucleic Acids from Cell and Matrix Debris		71
		4.4.4.1	Phenol Quality	71
		4.4.4.2	Inefficient Phenol Extraction and Removal	71
	4.4.5	Additional Purification of DNA		71
		4.4.5.1	Composition of Extraction Buffer	72
		4.4.5.2	Column Cleaning	72
		4.4.5.3	RNase Treatment of the Sample	72
	4.4.6	Precipitation and Concentration of DNA		74
		4.4.6.1	Volume and Temperature of Alcohol Used and Precipitation Times	74
		4.4.6.2	Concentration and Type of Salt	74
		4.4.6.3	Degraded DNA	74
		4.4.6.4	DNA Concentration	75
		4.4.6.5	Pellet Loss	75
		4.4.6.6	Pellet Incompletely Re-suspended	75
		4.4.6.7	Alcohol Precipitated Inhibitors	76

4.5	Automation of DNA Extraction	76
4.6	DNA Extraction Protocols	77
4.7	Summary	79
References		79

Chapter 5 DNA Quantification
Paul A. Heaton and Jacquie T. Keer

5.1	Introduction	83
5.2	Measurement of DNA Concentration Using Ultraviolet Spectroscopy	84
	5.2.1 Determining the Extinction Coefficient ε	84
	5.2.2 Practical Aspects of Measuring DNA Concentrations by UV Spectroscopy	85
	5.2.2.1 Calibration of the Spectrophotometer	85
	5.2.2.2 Cuvettes	85
	5.2.2.3 Sample Preparation	86
	5.2.2.4 Reference Blank	86
	5.2.2.5 Sample Dilution	86
	5.2.2.6 Light Source	86
	5.2.2.7 Presence of Contaminants	87
5.3	Determination of DNA Concentration by Fluorescence Spectroscopy	88
	5.3.1 Preparation of a Calibration Graph	88
	5.3.2 Practical Aspects of Measuring DNA Concentrations by Fluorescence Spectroscopy	89
	5.3.2.1 Sample Preparation	89
	5.3.2.2 Reference Blank	90
	5.3.2.3 DNA Standard	90
	5.3.2.4 Selecting the Dye	90
	5.3.2.5 Dye Concentration	90
	5.3.2.6 Microtitre Plates	90
	5.3.2.7 Measurement Conditions	90
	5.3.3 Fluorescent Dyes	91
	5.3.3.1 Ethidium Bromide	91
	5.3.3.2 PicoGreen®	92
	5.3.3.3 SYBR Dyes	92
	5.3.3.4 Hoechst 33258	92
5.4	Quantification Using the Polymerase Chain Reaction	93
5.5	Enzymatic Quantification of DNA	93
5.6	Primary Methods of DNA Quantification	94
	5.6.1 Gravimetric Analysis	95
	5.6.2 Isotope Dilution Mass Spectrometry for Oligonucleotide Quantification	95

	5.7 DNA Quantification by Constituent Phosphorus Determination	97
	5.8 Comparability of DNA Measurement Methods	98
	5.9 Summary	99
	References	99

Chapter 6 PCR: Factors Affecting Reliability and Validity
Charlotte L. Bailey, Lyndsey Birch and David G. McDowell

6.1	Introduction	101
	6.1.1 Real-time PCR	103
6.2	The Amplification Protocol	103
6.3	DNA Template	105
	6.3.1 Integrity	105
	6.3.2 Concentration	105
6.4	Reaction Components and Conditions Affecting Amplification and Reliability	106
	6.4.1 Reaction Buffer	106
	6.4.2 Magnesium Chloride	107
	6.4.3 Deoxynucleotide Triphosphates	107
	6.4.4 Water	107
	6.4.5 Primer Design and Target Selection	107
	6.4.5.1 Primer Design	107
	6.4.5.2 Target Selection	109
	6.4.6 Thermostable DNA Polymerases	109
	6.4.6.1 Factors Affecting Choice of Polymerase	110
	6.4.7 Hot-start Mechanisms	110
	6.4.7.1 Wax	112
	6.4.7.2 Antibody	112
	6.4.8 PCR Optimisation	112
	6.4.9 Inhibitors and Enhancers	113
6.5	Thermal Cycling	119
	6.5.1 Cycle set-up	119
	6.5.1.1 Denaturation	119
	6.5.1.2 Annealing	119
	6.5.1.3 Extension	119
	6.5.1.4 Cycle Number	120
	6.5.2 Thermal Cycler	120
	6.5.3 Temperature Control	120
	6.5.3.1 Block Control	121
	6.5.3.2 Reaction Control	121
	6.5.4 Ramp Rate	121
	6.5.5 Alternative Thermal Cyclers	122

6.6		Contamination Control	122
	6.6.1	Physical Laboratory Separation and Dedicated Equipment	123
	6.6.2	Pipettes	124
	6.6.3	Methods of Decontamination	124
		6.6.3.1 Uracil-N-glycosylase and dUTP	124
		6.6.3.2 Ultraviolet Light	128
		6.6.3.3 Chemical Decontamination	128
6.7		Post-PCR Analysis	128
6.8		Conclusions	128
References			129

Chapter 7 Quantitative Real-time PCR Analysis
Jacquie T. Keer

7.1		Introduction	132
7.2		Approaches to Product Detection	133
	7.2.1	The 5′ Nuclease Assay	134
	7.2.2	Molecular Beacons™	135
	7.2.3	Hybridisation Probes	136
	7.2.4	Scorpion™ Primers	138
	7.2.5	Plexor™ Primer Technology	138
	7.2.6	Melting Curve Analysis	139
	7.2.7	Choice of Fluorophores	140
7.3		Range of Instruments	141
7.4		Practical Aspects of qPCR Analysis	144
	7.4.1	Assay Design	144
		7.4.1.1 Target Sequence	145
		7.4.1.2 Probe and Primer Design	145
	7.4.2	PCR Master Mix	146
		7.4.2.1 Magnesium Chloride	146
		7.4.2.2 DNA Polymerase	147
	7.4.3	Cycling Conditions	147
	7.4.4	Primer and Probe Optimisation	149
	7.4.5	Target Level	150
	7.4.6	Contamination Control	152
	7.4.7	Experimental Design	153
		7.4.7.1 Use of Controls	153
		7.4.7.2 Level of Replication	153
		7.4.7.3 Randomisation	154
	7.4.8	Data Analysis	154
		7.4.8.1 Basic Mathematics of PCR Amplification	155
		7.4.8.2 Data Normalisation	155

		7.4.8.3	Routes to Determining Amplification Efficiency	156
		7.4.8.4	Outlier Identification	156
	7.4.9	Validation		156
7.5	Quantification of Low Levels of Target Analyte			157
	7.5.1	Level of Variability		158
		7.5.1.1	Sample Handling	158
		7.5.1.2	Amplification Cycles	159
		7.5.1.3	Replication Level	159
		7.5.1.4	Data Handling	159
7.6	Standards and Comparability			161
	7.6.1	Quantitative Standards		161
		7.6.1.1	Instrument Calibration	162
		7.6.1.2	Comparability	162
		7.6.1.3	Measurement Uncertainty	163
7.7	Summary			163
References				164

Chapter 8 Multiplex PCR and Whole Genome Amplification
Lyndsey Birch, Charlotte L. Bailey and Morten T. Anderson

8.1	Introduction to Multiplex PCR		167
	8.1.1	Number of Targets Amplified During Multiplex PCR	168
8.2	Design and Optimisation of mPCR		168
	8.2.1	Design Strategy	168
	8.2.2	Amplification Target	169
	8.2.3	Primer Positioning	169
	8.2.4	Primer Design	170
	8.2.5	Standardisation of Oligonucleotide T_m	171
		8.2.5.1 Base Analogues	171
		8.2.5.2 Peptide Nucleic Acid (PNA)	171
		8.2.5.3 Locked Nucleic Acid (LNA)	171
	8.2.6	Optimisation	174
		8.2.6.1 Initial Assay Development	174
		8.2.6.2 Reaction Components	174
		8.2.6.3 Cycling Parameters	175
	8.2.7	Overcoming Mis-Priming Events	176
	8.2.8	Specificity	176
	8.2.9	Untemplated Nucleotide Addition	177
8.3	Detection Strategies		177
8.4	Applications of mPCR		177
8.5	Advantages and Disadvantages of mPCR		177
8.6	Introduction to WGA		179

	8.7	WGA Methodologies	179
		8.7.1 Degenerate Oligonucleotide Primed PCR	179
		8.7.2 Primer Extension Pre-amplification	180
		8.7.3 Improved Primer Extension Pre-amplification	180
		8.7.4 Multiple Displacement Amplification	180
		8.7.5 Ligation-mediated PCR	181
		8.7.6 T7-based Linear Amplification of DNA	181
	8.8	WGA Applications and Characteristics	182
	Acknowledgements	184
	References	184

Chapter 9   Procedures for Quality Control of RNA Samples for Use in Quantitative Reverse Transcription PCR
Tania Nolan and Stephen Bustin

9.1	Introduction	189
9.2	RNA Extraction Approaches	189
	9.2.1 Freezing	189
	9.2.2 Sulfate	190
	9.2.3 Guanidinium Isothiocyanate	190
	9.2.4 Phenol	190
	9.2.5 Additional Purification	190
	9.2.6 Extraction from Archival Tissue Samples	191
9.3	RNA Quality	192
	9.3.1 RNA Integrity Number	193
	9.3.2 Spectrophotometric Measurement	194
	9.3.3 Presence of Inhibitors	194
9.4	RNA Quantification	199
	9.4.1 Significance of Quantification	199
	9.4.2 Methods of Quantification	201
9.5	Effect of RT Experimental Design on qPCR Data	201
9.6	Conclusion	204
References	205

Chapter 10   Microarrays
Sally L. Hopkins and Charlotte L. Bailey

10.1	Introduction	208
	10.1.1 What are Microarrays?	208
	10.1.2 A Note about Nomenclature	210
	10.1.3 Types of Microarrays	210
	10.1.3.1 Applications of DNA Microarrays	212
	10.1.3.2 Impact of Applications	213
10.2	Technology Status	214
	10.2.1 Current Problems	214
	10.2.2 Controlling Experimental Uncertainties	216

			10.2.2.1 Experimental Design	217

              10.2.2.1 Experimental Design   217
              10.2.2.2 Microarray Layout and Content   219
              10.2.2.3 Target Quality   220
              10.2.2.4 Array Handling and Hybridisation   221
              10.2.2.5 Gene List Files   222
              10.2.2.6 Image Acquisition and Processing   223
              10.2.2.7 Normalisation   224
              10.2.2.8 Critical Data Assessment   224
              10.2.2.9 Drawing Biological Conclusions   225
              10.2.2.10 Data Management   225
        10.2.3 Technology Solutions   226
    10.3 Current Commercial Microarray Quality Controls   226
        10.3.1 Printing Controls   227
        10.3.2 Universal Reference RNA   228
        10.3.3 Spike-in Controls   229
    10.4 Microarray Standardisation Initiatives   229
        10.4.1 Microarray Gene Expression Data Society (MGED)   230
        10.4.2 External RNA Control Consortium (ERCC)   230
        10.4.3 Microarray Quality Control (MAQC) Project   231
        10.4.4 Association of Biomolecular Research Facilities (ABRF) Microarray Research Group (MARG)   231
        10.4.5 Measurements for Biotechnology (MfB) Programme   232
              10.4.5.1 Specificity Standards and Performance Indicators   232
              10.4.5.2 Comparability of Gene Measurements   234
              10.4.5.3 Quality Metrics/Increasing Confidence in Toxicogenomic Measurement   234
              10.4.5.4 Standard Units to Measure Gene Expression   235
    10.5 Summary   235
    References   235

Subject Index   240

Abbreviations

A	adenine
A_x	absorbance at x nm
ABRF	Association of Biomolecular Research Facilities
AFLP	amplified fragment length polymorphism
BBSRC	Biotechnology and Biological Sciences Research Council
BIPM	International Bureau of Weights and Measures
bp	base pair
BSA	bovine serum albumin
BSI	British Standards Institute
c7dGTP	7-deaza-2′-deoxyguanosine triphosphate
C	cytosine
cDNA	complementary DNA
CGH	comparative genome hybridisation
CRM	certified reference material
Ct	cycle threshold
CTAB	cetyltrimethylammonium bromide
CV	coefficient of variation
dATP	deoxyadenosine triphosphate
dCTP	deoxycytidine triphosphate
dGTP	deoxyguanosine triphosphate
DEFRA	Department of Environment, Food and Rural Affairs
DMSO	dimethyl sulfoxide
DNA	deoxyribonucleic acid
DNase	deoxyribonuclease
dNTP	deoxyribonucleotide triphosphate
DOP-PCR	degenerate oligonucleotide primed PCR
dsDNA	double-stranded DNA
DTI	Department of Trade and Industry
dTTP	deoxythymidine triphosphate
dUTP	deoxyuridine triphosphate
EBI	European Bioinformatics Institute
EC	European Community
EDTA	ethylenediaminetetraacetic acid

EGTA	ethyleneglycol-bis(ß-aminoethyl ether)tetraacetic acid
EQA	external quality assessment
ERCC	External RNA Control Consortium
ESI	electrospray ionisation
EtBr	ethidium bromide
FDA	Federal Drug Administration
FRET	Förster Resonance Energy Transfer
FSA	Food Standards Agency
G	guanine
GAL	gene array list
GITC	guanidinium isothiocyanate
GLP	Good Laboratory Practice
GLPMA	GLP Monitoring Authority
GM(O)	genetically modified organism
HSE	Health and Safety Executive
HMW	high molecular weight
ICP	inductively coupled plasma
ICP-OES	inductively coupled plasma optical emission spectroscopy
IDMS	isotope dilution mass spectrometry
IEC	International Electrotechnical Commission
I-PEP	improved primer extension pre-amplification
ISO	International Organization for Standardization
IUPAC	International Union of Pure and Applied Chemistry
kb	kilobase
LIMS	Laboratory Information Management System
LMP	ligation-mediated PCR
LNA	locked nucleic acid
LoD	limit of detection
LOH	loss of heterozygosity
LoQ	limit of quantification
MALDI-ToF	matrix assisted laser desorption ionisation time of flight
MDA	multiple displacement amplification
MGB	minor groove binder
MGED	Microarray Gene Expression Data Society
MHRA	Medicines and Healthcare Products Regulatory Agency
mPCR	multiplex PCR
mRNA	messenger RNA
MS	mass spectrometry
MWM	molecular weight marker
n/s	nucleotides per second
NCBI	National Center for Biotechnology
NIST	National Institute of Standards and Technology
NERC	Natural Environmental Research Council
NP40	nonidet P40
OD	optical density
OECD	Organisation for Economic Co-operation and Development

Abbreviations

OES	optical emission spectroscopy
PAGE	polyacrylamide gel electrophoresis
PBS	phosphate buffered saline
PCR	polymerase chain reaction
PEG	polyethylene glycol
PEP	primer extension pre-amplification
PFGE	pulsed field gel electrophoresis
PNA	peptide nucleic acid
PVP	polyvinylpyrrolidone
PT	proficiency testing
QA	quality assurance
QC	quality control
QP	quality procedure
qPCR	quantitative PCR
qRT-PCR	quantitative reverse transcription PCR
RAPD	randomly amplified polymorphic DNA
RCA	rolling circle amplification
RFLP	restriction fragment length polymorphism
RIN	RNA integrity number
RM	reference material
RNA	ribonucleic acid
RNase	ribonuclease
rRNA	ribosomal RNA
RT-PCR	reverse transcriptase PCR
SBE	single base extension
SCOMP	single cell comparative genomic hybridisation
SDS	sodium dodecyl sulfate
SNP	single nucleotide polymorphism
SOP	standard operating procedure
SSC	salt sodium citrate
SSCP	single-strand conformation polymorphism
ssDNA	single-stranded DNA
STR	short tandem repeat
T	thymine
Taq	*Thermus aquaticus*
TE	tris-EDTA buffer
T_m	melting temperature
TLAD	T7-based linear amplification of DNA
TMAC	tetramethylammonium chloride
Tris	tris(hydroxymethyl)aminomethane
U	uracil
UKAP	United Kingdom Analytical Partnership
UKAS	United Kingdom Accreditation Service
UNG	uracil-N-glycosylase
UV	ultraviolet
VAM	Valid Analytical Measurements

WI	working instruction
WGA	whole genome amplification

Weights, volumes and concentrations

M	molar
mol	mole
g	gram
L	litre
v/v, vol/vol	volume per volume
w/v	weight per volume

Prefixes

m	milli (10^{-3})
μ	micro (10^{-6})
n	nano (10^{-9})
p	pico (10^{-12})
f	femto (10^{-15})
a	atto (10^{-18})
z	zepto (10^{-21})

Time

s	second
h	hour
min	minute

Acknowledgements

We wish to thank all of our colleagues who have contributed to this manual. Our particular thanks go both to the authors of individual chapters, who are therefore acknowledged directly, and to Claire English, Shivani Mehta, Carol Donald, Gavin Nixon and Hernan Valdivia of LGC for their contribution of additional experimental data in support of the issues raised. Dr Liz Prichard, Dr Alison Woolford, Dr Carole Foy, Dr Malcolm Burns and Dr David French of LGC have kindly assisted in preparation of the manuscript by critical reading and feedback on the material.

The work described was supported under contract with the Department of Trade and Industry as part of the National Measurement System Measurements for Biotechnology Programme 2004–2007.

Jacquie T. Keer
Lyndsey Birch

CHAPTER 1
Valid Analytical Molecular Biology: The Challenge

JACQUIE T. KEER

LGC, Queens Road, Teddington, TW11 0LY

1.1 Introduction

The last decade has seen a rapid increase in the pace of technological advancement and in the uptake of DNA analysis for a range of applications. The increased use of DNA as an analyte reflects its uniform presence in almost all cells of most organisms. In addition the greater stability of DNA, compared to RNA or protein molecules, is ideal for analysis of highly processed or aged samples.

Technical innovations include the development of more sensitive, quantitative, high-throughput and massively parallel analyses, all generating new applications and commercial opportunities and covering a wide range of uses. The complete DNA sequence of many genomes has been determined, opening the way for a plethora of new applications, including directed drug discovery and personalised genetic diagnostics and treatment. Forensic analysis, food testing and agriculture are just a few of the many other areas where DNA technology is being adopted, with concomitant changes in regulation and procedures. It is clear that there are significant advantages in using molecular methods, including reduced detection limits, greater speed and scale, lower cost and improved specificity. The potential of novel genetic diagnostic methods, directed drug discovery routes and the increased throughput of massively parallel array-based analyses are strong drivers for even greater uptake of this technology. However, to exploit fully the potential of these developments and remove barriers to wider uptake, there is a need to ensure that molecular analytical methods are reliable, consistent and fit for purpose, in order to avoid the use of biased or flawed techniques and resultant loss of confidence in the techniques.

The majority of technological development occurs in academic or medical research environments, where the main priority is innovation. Consequently little consideration is given to the more routine applicability, reliability and reproducibility of methods, particularly in the early stages of development. Despite evaluation of method performance characteristics and method validation being a prerequisite for the successful move of techniques from the research laboratory to the analytical laboratory, there is resistance to such formal evaluation in some sectors. There are also practical barriers to assessment of method performance, including the lack of reference materials which are necessary for the critical comparison of analytical approaches and the paucity of performance standards in the wider analytical community, as most regulation of analysis is carried out in-house. However, in the light of growing commercial and clinical application, consideration is increasingly being given to the reliability of the technology being used.

Although large volumes of analytical data may be produced from poorly applied methods, generation of dependable results usually requires careful and considered planning and validation. The aim of any experiment is to produce reliable results, and to avoid the need to repeat the analysis because of problems with the reagents, method or equipment used. Consistently 'getting it right first time' depends on a number of factors, including provision of a controlled laboratory environment with calibrated and regularly maintained instruments, use of an effective experimental design and performance of the work by an analyst with sufficient training and experience to correctly perform the method and interpret the result (Figure 1.1). Although it is difficult to estimate the actual cost of poor laboratory practice in wasted time and reagents, the benefits in avoiding repeating work are very clear.

This manual aims to introduce and address quality assurance and validation issues that arise in the application of DNA technology, and to provide a basis for the development of validated methods and experimental good practice. Specifically, Chapters 2 and 3 cover the benefits of formal laboratory management systems and method validation. The remaining chapters in the manual provide information on a range of commonly used techniques, from the initial extraction of DNA from analytical samples and quantification of the amount of DNA present, to a range of downstream processes including various forms of polymerase chain reaction (PCR) amplification and microarray-based analysis.

Analytical laboratories should work to produce quality analytical data, and reading the information presented here should provide a firm foundation for good experimental practice.

1.2 The Analytical Process

1.2.1 Analytical Requirements

Analysis is usually initiated by a 'customer', who can be a private individual or company, public organisation, research funding body or law enforcement

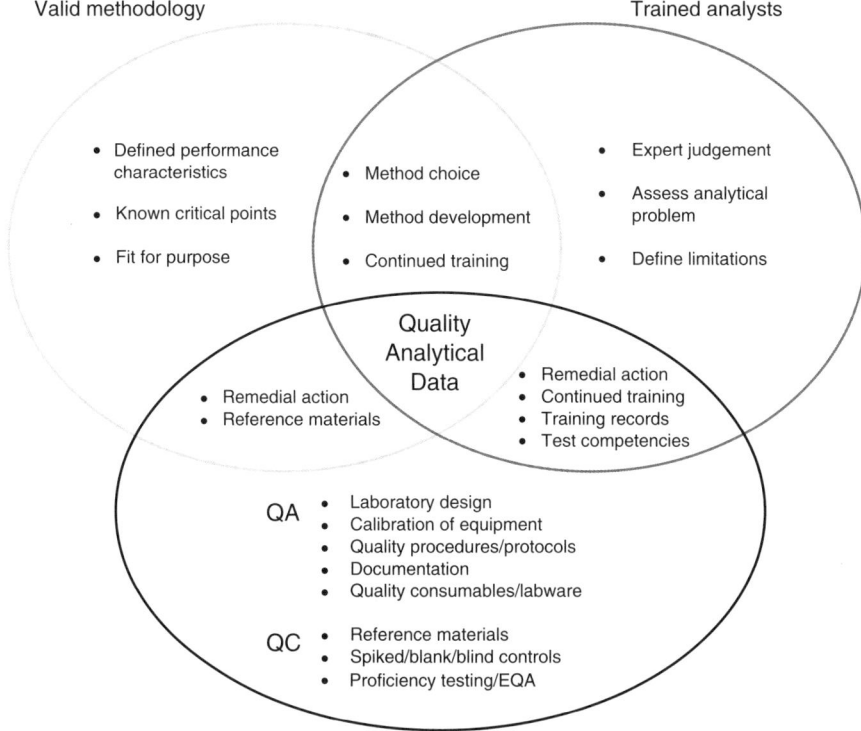

Figure 1.1 Schematic diagram showing the factors within the laboratory that contribute to the production of reliable data.

agency such as a police force or trading standards office. The results that are produced are usually required for a specific purpose, often as an independent source of information in order to gauge a situation, interpret evidence, determine whether action is required or to ascertain whether certain regulations are being adhered to. Increasingly, some indication of the level of confidence that can be placed in the result is also required, allowing the results of the experiment to be used or interpreted appropriately.

1.2.2 Stages in the Analytical Process

In undertaking an experiment or analysis to address a specific question, a complex procedure is undertaken, beginning with the initial researching of the questions and specific analytical requirements and ending with the interpretation of the analytical data produced and the reporting of results and conclusions. To ensure the process is efficient, careful planning of the work is required. Good experimental design, trained staff and use of suitable methods, equipment, standards and samples can save time in ensuring that sufficient and

Table 1.1 Stages of the analytical process.

Define the analytical enquiry	• Define type of data required (qualitative/quantitative) • Define use of the data and confidence required in the result • Define required performance of the method
Assess the sample	• Determine nature of the sample • Define storage, transport and preparation requirements • Use appropriate sampling procedures • Ensure trackability through unique sample ID system
Establish constraints	• Establish time available for analysis • Identify equipment and personnel resources available • Understand any financial limitations • Determine if there are special safety considerations • Is the analysis feasible?
Define technical approach	• Identify suitable techniques based on analytical requirement • Match analytical performance to requirements
Select or develop method	• Select suitable method from published literature, or commercial kits • If none suitable, develop in-house method • Finally, prepare a draft protocol
Validate method	• Ensure method performance meets analytical requirements • Demonstrate the method produces appropriate data • Document the suitability of the method
Apply the validated method	• Analyse the samples using the selected, validated approach • Use appropriate controls to enable confident interpretation of results
Interpret and report the data	• Interpretation of the data will depend on the results from QC materials included in the assay as well as test samples themselves • Any limitations of the method should be included in conclusions and interpretations of results

reliable results are produced first time. A flawed approach may produce experimentally valid data that do not directly address the enquiry, or insufficient data for confident interpretation. Incorrect sample collection or storage could produce erratic results even when a valid method is applied. In addition, use of uncalibrated equipment could generate biased results that do not allow correct judgement of the actual situation. An overview of the stages to consider when planning the experimental process is outlined in Table 1.1.

1.3 Principles Underpinning Reliable Measurement

A series of six principles to underpin good experimental practice has been developed, known as the Valid Analytical Measurement (VAM) principles.

Although primarily directed towards chemical analysis, the principles are generic and fully applicable to biological measurement performed in both research and more routine laboratory environments. The described approach requires support and implementation at both a technical and management level, and the ethos needs to be understood and supported by all staff in the laboratory.

1.3.1 Understand the Experimental Requirements

Experiments or measurements are generally undertaken to answer a specific question or to provide a solution to a problem. Understanding the full nature of the enquiry enables an experimental approach to be developed to produce sufficient data to fully answer the question.

1.3.2 Use Methods and Equipment which are Fit for the Intended Purpose

Consistent production of reliable data requires that the methods, instruments, reagents and software used in an analysis have been tested and shown to perform as expected. Further information on fulfilling these requirements is given in Chapters 2 and 3.

1.3.3 Staff Undertaking Analysis Should be Both Qualified and Competent to Undertake the Task

To ensure that methods and equipment are used correctly, appropriate levels of staff training and support are required. Formal management schemes address the continued training and assessment of staff (Chapter 2), and even in laboratories where no formal system is in place, some level of training is advisable to avoid time wasted in repeating experimental analyses and the cost of equipment damage through misuse.

1.3.4 Regular Independent Assessment of Laboratory Performance

Independent assessment usually takes the form of proficiency testing (PT) or external quality assessment (EQA) schemes, where samples are distributed to participating laboratories for analysis. The results are returned and analysed by the scheme organisers, and a report detailing the performance of the participants is produced, usually without identifying individual participants. Such external assessments of performance are useful to confirm that procedures are producing acceptable in-house results, and results can be compared to those produced in peer laboratories.

1.3.5 Analytical Consistency

A primary aim of any researcher or analyst is consistently to produce reliable and valid results. The use of well-defined samples or certified reference materials (CRMs) can be used on a regular basis to demonstrate the consistent quality of measurements within a laboratory. In biological analyses few reference materials are available, but the use of previously characterised samples can be used to monitor performance over time.

1.3.6 Quality Control and Quality Assurance Framework

Formal management systems (Chapter 2) specify the need for laboratory management systems, including the use of trained staff, calibrated equipment, quality protocols and valid methodologies. This is the quality assurance (QA) framework, which can assist in preventing errors by ensuring the laboratory and analytical environment is fit for purpose. Quality control (QC) measures are used in parallel with QA systems, and confirm the quality of data obtained by the use of control samples and continual monitoring of performance.

1.4 Challenges to Measurement Quality

Despite the establishment of good measurement principles, many technical challenges remain, arising from a number of factors including the variety of available methods and platforms, the pace of technological development, lack of certified reference materials to establish comparability between approaches and few accessible EQA or PT systems to evaluate comparability between laboratories. In addition to the practical challenges there is also a number of administrative and management issues including pressure to publish results regularly, often high levels of staff turnover and lack of funds and resources for QC and QA activities.

The analysis of real samples often provides a further challenge, as low concentrations or inhomogeneous distribution of targets may pose problems. Difficult samples may originate from a variety of sectoral applications such as forensic, food or environment, where the DNA analyte may be in association with an organic matrix, for example, a blood stain on cotton fibre, bacterial species in milk or genetically modified (GM) soya in processed food products. A number of common problems that arise in the application of DNA technologies are considered here.

1.4.1 Low Concentration of Analyte Compared to Matrix

The need to detect, identify and/or quantify very low levels of the target in a large amount of sample matrix for various applications has led to the development of sophisticated DNA extraction and amplification methodologies to selectively isolate and concentrate the analyte of interest. Examples include low-level detection of environmental and food pathogens, non-invasive prenatal diagnostic methods and the quantification of DNA contaminants in biopharmaceuticals.

1.4.2 Complex Matrices

Target analytes present in complex chemical or biological matrices can make sampling and DNA extraction a difficult undertaking. Challenging matrix components include naturally occurring secondary compounds which can interfere with enzyme activity and can cause total inhibition of biological reactions such as PCR and restriction enzyme digests. Certain components of biological samples, such as haem and urea, may also affect the analysis. There may also be difficulties in physically separating the analyte from the matrix, as clumping and adhesion can make uniform sampling and efficient DNA extraction difficult.

1.4.3 DNA Degradation

In some instances samples may be subjected to harsh environmental, transport or storage conditions that can damage the analytes significantly. Poor conditions include industrial processing such as freezing, dyeing, heating, grinding, tanning, drying and forms of weathering such as those caused by the sun or rain. Ageing of a sample can also cause physical degradation of the DNA analyte, and in such instances the use of short DNA targets can enable even highly degraded materials to yield results. However, it is important to use calibrators and controls that are in an equivalent physical state to the test sample, as otherwise the results of the analysis may be affected by any disparate performance of the intact and degraded materials in the experiment.

1.4.4 Biological Contamination of the Sample

Often the test sample has been contaminated before arrival at the testing laboratory, so nucleic acids from a variety of sources may be present. The contamination may be due to environmental insult (for example, bacterial or fungal contamination), or may inherently be a mixture of target and non-target material (such as food samples containing a proportion of genetically modified ingredients). A further problem may arise if the contaminating material contains chemicals or enzymes that are able to damage DNA within the sample, such as low pH fruit juices or DNases from contaminating micro-organisms. In this situation it is not always possible to utilise controls in a similar physical state to the test samples, and so the likely effect of the contaminant on the results should be taken into account when interpreting the results of the experiment.

1.4.5 Degradation of Matrix Components

Various components within an analytical matrix can sometimes produce breakdown products, such as polyphenols, that cause the degradation of nucleic acids. In such circumstances, some assessment of the level of DNA degradation is helpful to enable the results of the test to be interpreted correctly. Tests of the effect of known contaminants may also be performed, using appropriately spiked samples as controls.

1.4.6 Limited Availability of the Sample

A sample may be limited because the sample represents a unique moment in time (for example, a particular stage in disease progression) or is limited by quantity (such as a scene-of-crime swab or patient biopsy). Samples that are difficult to replace require extreme care in processing and analysis, as by definition the analysis cannot easily be repeated.

1.4.7 Lack of Suitable Controls

One of the main challenges in molecular analysis is the lack of certified reference materials (CRMs), which are a key component in validation and QA procedures. The majority of physical and chemical measurements are underpinned by suitable standards but, because of the complexity of the analytes and the wide range of materials under test, such standards are not available for most biological applications. In addition, there are very few characterised reference samples that can be employed to ensure the accurate calibration of equipment, the correct handling of samples or applicability of methodologies. Alternatives to the use of CRMs include comparison of the results from several techniques, or analysis of the same material by several laboratories, followed by comparison of the results. However, as there are no materials for which a 'true value' is known, objective assessment of method, equipment and analyst performance is not straightforward.

1.5 Focus on Data Quality

As mentioned already, most QA and QC activity takes place in-house, and may sometimes be compromised because of resource constraints. An exception is in analyses where the results may be reported and used in a court of law, such as short tandem repeat (STR) forensic profiling and the quantitative determination of GM ingredients in foodstuffs. In such cases, the validity of the analytical result must be demonstrated absolutely, and is subject to stringent questioning and challenge by defence lawyers. The presence of an equivalent pressure is not always evident in other areas of analytical molecular biology. The quality of data is reliant on the professionalism of the analytical laboratory and the analysts involved, requiring continual questioning and re-evaluation of the analytical approach, procedure, staff capabilities and applicability of the test. Individuals within the laboratory need to look beyond the data that are produced by the instrument or method to the wider experimental context in order to correctly interpret results. For example many quantitative PCR (qPCR) instruments provide quantification results with an apparent accuracy of several decimal places. However, consideration of the likely errors introduced into the process through the various stages including preparation of standards and reaction set-up indicates that reporting results with this level of precision may be misleading.

Exercising critical judgement in laboratory set-up, experimental design and practice and in interpretation of results is central to ensuring consistent production of reliable data and, most importantly, is in your hands.

Acknowledgements

The author would like to thank Ginny Saunders for information and illustrations, and Lyndsey Birch for many helpful discussions.

CHAPTER 2
Quality in the Analytical Molecular Biology Laboratory

SALLY L. HOPKINS

LGC, Queens Road, Teddington, TW11 0LY

2.1 Introduction

Definition of terms:

In this chapter the terms traceability and metrological traceability will be used. It is important for the reader to understand the meanings of these terms as used here:

- Traceability (also known as trackability in the chemical industry) is used to mean the traceability of an entire experimental procedure, through reagent batch numbers, sample identifier codes and unique file names;
- Metrological traceability is used to mean the ability to trace the value of a result from experimental processes back through an unbroken chain of comparison, each with a stated uncertainty, to national or international standards.

The need for valid practices to produce traceable and robust data of acceptable quality cannot be disputed in any analytical environment; however the route to consistently obtaining such analytical data is not necessarily a clear and straightforward path. This chapter will attempt to demonstrate how implementation of quality procedures can support the ultimate goal of the analyst, namely getting the analysis right first time and every time.

Defining quality is difficult and there are many different definitions. One of the most common themes is that quality is about matching a product or service with the requirements of the customer.

Most laboratories probably have some form of management system, even if it is not formally accredited or certified, as without one work may get out of control very quickly. Currently, however, there is increasing emphasis on formal management systems, certification, compliance and accreditation, assessed by nationally or internationally approved third parties. These formal systems cost much in time and money to implement, so why should we bother implementing them at all?

A formal management system is an internationally recognised standard, which is acknowledged and mutually accepted by customers and other organisations around the world, identifying that work is of a consistent standard. It provides a means of structuring the processes and procedures used in an organisation, making it easier to identify potential risk areas and correct problems. By focusing on the performance of the organisation and the competency of the staff, documenting all data and keeping records of processes and procedures, the volume of errors reduces, and proof of performance is available and defendable to third parties. This means that the management system could help to improve the status of a laboratory amongst other companies and customers, as well as saving money by reducing the number of repeat measurements being carried out. Both these benefits provide an organisation with a competitive edge.

There is also an increasing drive from customers to find credible laboratories, which can demonstrate the production of quality data. Customers are now frequently asking for proof of competence in the form of certification, accreditation or compliance from approved third parties. Customer perceptions were explored in a UKAP-funded survey, in which private sector customers quoted appropriate accreditation as the most important attribute for a potential supplier.[1] As well as this, some UK funding bodies are asking for evidence of best practices and management systems before handing over research grants.[2]

There is a number of possible consequences in not following a management system, including potential lack of control over processes and procedures, inconsistency of analytical approach and lack of measurement and process traceability, all leading to reduced confidence in the data produced by the laboratory. The organisation may not be managing its risks effectively and may, therefore, be open to legal challenge regarding the quality of results or other work. It is also harder to demonstrate independently the quality of work carried out, without nationally or internationally recognised third-party confirmation of a laboratory' quality status. Ultimately, if an organisation does not have independent recognition, in terms of quality, then it risks losing existing customers and may find it difficult to gain new ones.

The aim of the following sections is to provide a brief guide through some of the fundamentals of management systems and give some guidance on their implementation. This may help a laboratory decide which standard best suits its business needs.

2.2 Management Systems

It is of paramount importance to an analytical laboratory that a management framework is set up in order to ensure that all analysis is performed with sufficient assurance of quality. Quality assurance and quality control may be defined as follows:

- *Quality assurance (QA)*
  A planned set of activities designed to ensure that the quality control programme is effective. It is the overarching system, which plans and documents the processes involved in ensuring quality output.
- *Quality control (QC)*
  A planned system of activities designed to provide a quality product. These activities are planned in the QA system.

A management system contains policies, procedures and instructions that set out how things will be done and will also help demonstrate performance of correct procedure, covering many aspects of administration and laboratory organisation. Some such aspects are:

- Quality policy statement;
- General organisation;
- Roles and responsibilities of staff;
- Document control;
- Quality manual.

A range of more specific topics may also be included:

- Laboratory environment;
- Security;
- Facilities and equipment used for testing/calibration;
- Employment of suitable staff and their training;
- Procedures relating to sample handling and control;
- Test methods and procedures;
- Policy on subcontracting of work and reporting results;
- Use of valid methodologies and QC measures.

An auditing and review policy and procedure must also be in place to ensure that all documentation and procedures are completed as required. The QA system is distinct from the QC process. QA is about demonstrating that the Quality Control process is effective, and how it is maintained under control. The QC process describes the day-to-day activities which are carried out to provide a series of checks on the product. This will be described in more detail in Section 2.4.7.

2.3 Internationally Recognised Assessed Standards

Many organisations operate a highly structured management system similar to that set out in Figure 2.1.

Figure 2.1 Schematic diagram of a typical management system structure.

- **Certification** to a standard is obtained when a specified third party issues a statement, based on a decision following review, that fulfilment of specified requirements related to products, processes, systems or persons has been demonstrated.[3]
- **Accreditation** is achieved when a third party issues a statement, based on a decision following review, that competence to carry out a task has been demonstrated.[3]

There are four commonly used international standards for laboratories, all of which complement each other. These are ISO 9001:2000, ISO/IEC 17025:2005, ISO 15189:2003 and the Principles of Good Laboratory Practice (GLP). ISO (the International Organization for Standardization) and IEC (the International Electrotechnical Commission) form the specialised system for worldwide standardisation, while GLP has been developed by the Organisation for Economic Co-operation and Development (OECD). The type of analysis the laboratory carries out will largely govern the standard(s) that it adopts.

As well as the four international standards above, the analytical molecular biology laboratory may also need to be aware of recommendations highlighted by funding bodies. Increasingly, funding bodies are asking for assurance of the quality of data produced. In the UK the Joint Code of Practice for Research[2] was issued in 2003 by the Department of Environment, Food and Rural Affairs (DEFRA), the Food Standards Agency (FSA), the Biotechnology and Biological Sciences Research Council (BBSRC) and the Natural Environmental Research Council (NERC). The code lays out guidelines for the quality of the research process and the quality of the science carried out, to ensure their contractors are using 'best scientific practice'. This code will be discussed along with the internationally recognised standards, to give laboratories an indication of what funding bodies will be looking for in the future.

2.3.1 ISO 9001:2000 Quality Management Systems – Requirements

ISO 9001:2000[4] is part of a family of standards:

- ISO 9000:2000 – Quality Management Systems: Concepts and Vocabulary;
- ISO 9001:2000 – Quality Management Systems: Requirements;
- ISO 9004:2000 – Quality Management Systems: Guidance for Performance Improvement.

ISO 9001:2000 is the standard within this family to which organisations will be assessed and awarded certification. It is this quality management standard that is commonly used by organisations manufacturing or supplying products or services in the UK and across the world. This standard replaced the ISO 9001:1994 series of standards, and places more emphasis on customer requirements, satisfaction and continual improvement.

This standard is generic in terms of its requirements and can, therefore, be applied to all types of organisations. It is concerned with controlling processes, as a way of meeting customer requirements and requires continuous improvement, demonstrating that quality is not a static activity. However, the standard is not prescriptive in terms of the technical requirements of the organisation or laboratory.

The certification process involves an organisation registering with an accredited certification body, who will usually discuss the organisation's needs and any fees which will be payable. Once the formal application and registration fees have been received it is usual for a pre-assessment visit to be carried out to review the organisation's documentation and further discuss the implementation of the standard. Once the organisation has put in place all that was suggested in the pre-assessment, an assessor will visit the organisation and finally assess their management system against the requirements of ISO 9001:2000. If the management system complies with these requirements, a formal confirmation and a certificate is issued which may then be used to demonstrate the achieved quality standard to other customers and organisations.

Once certification has been achieved, the certification body will re-assess the organisation at regular intervals, to ensure the management system is maintained at a satisfactory level. For example, the British Standards Institute (BSI) will visit at regular intervals (at least every year) to facilitate improvement as well as checking that the requirements of the standard are still being met.[5] The assessors will report to the organisation any non-conformity against the standard, as well as suggestions for improving the system, which may not be classed as non-conformities. Non-conformities must be corrected within a specified time period and the certifying body will require an action plan to be provided.

2.3.2 ISO/IEC 17025:2005 General Requirements for the Competence of Testing and Calibration Laboratories

ISO/IEC 17025:2005 is the current internationally accepted standard for the accreditation of testing and calibration laboratories.[6] This standard replaces the ISO/IEC 17025:1999 requirements, which in turn replaced ISO Guide 25 and many national standards. There are two main sections to this standard; one covering the management requirements and the other covering the technical requirements. ISO/IEC 17025 was updated in 2005 in order to bring it into alignment with ISO 9001:2000 (see Section 2.3.1). The main change in the management section is the specific requirement for continual improvement and communication with the customer to ensure their requirements are fully met. Throughout this chapter references to ISO/IEC 17025 refer to the 2005 standard.

ISO/IEC 17025:2005 specifies general requirements for demonstrating competence to carry out tests or calibrations and covers the use of standard, non-standard and laboratory-developed methods. This standard contains the management requirements of ISO 9001:2000, as well as many more technical specifications, including the laboratory environment, method validation, method uncertainty, metrological traceability and sampling. The standard is designed to assess and demonstrate competence for the tests or calibrations being carried out. If a laboratory is accredited to ISO/IEC 17025:2005 then it is stated that the management system, for these accredited activities, also meets the principles of ISO 9001:2000, as stated in the joint ISO-ILAC-IAF communiqué.

The majority of national accreditation bodies across the world accept the standard, and many of these bodies have mutual recognition agreements to accept the accreditation made and granted in other countries. In the UK, the competent body is the United Kingdom Accreditation Service (UKAS). Accreditation to ISO/IEC 17025:2005 is sought from UKAS for specific tests, in terms of the scope of a particular method, such as a particular analyte, matrix and instrument platform. Guidance and information for those seeking accreditation, including notes to aid implementation, application forms and fee schedule, are freely available from the UKAS website.[7]

On receipt of the relevant application forms and the application fee, UKAS will assign an assessment manager, who will arrange for a pre-assessment to be carried out. This visit addresses the scope of the accreditation sought. An initial assessment visit is then carried out to formally assess the applicant against ISO/IEC 17025:2005. The assessment manager and technical assessors, relevant to the scope of accreditation being sought, perform the assessment. Any non-conformities highlighted during the visit are notified to the applicant in writing. The applicant will be granted accreditation once the non-conformities are cleared to the satisfaction of UKAS. The maintenance of accreditation is confirmed by annual surveillance visits, the first of which takes place six months after the granting of accreditation. A full re-assessment is carried out every four years. At each of these stages the assessment manager will make a quotation for the charges due. Where available, participation in relevant proficiency testing (PT) schemes is also expected of ISO/IEC 17025:2005 accredited laboratories, and this will add to the cost of implementing the standard.[6]

2.3.3 ISO 15189:2003 Medical Laboratories – Particular Requirements for Quality and Competence

This standard matches the quality management requirements outlined in ISO 9001:2000. ISO 15189:2003[8] also covers most of the requirements of ISO/IEC 17025:2005, although some aspects are covered in lesser detail, such as method validation, and some in more detail, such as sampling. In contrast to ISO/IEC 17025:2005, the focus is on patient outcome, without downgrading the need for accuracy of measurement. ISO 15189:2003 emphasises not only the quality of measurement, but also the total service of a medical laboratory, including aspects like consultation and cost effectiveness. The standard also highlights important pre-and post-investigative issues and addresses ethics and the informational needs of the medical laboratory. ISO 15189:2003 does lack detail in terms of traceability and measurement uncertainty; this standard is currently under revision as ISO/FDIS 15189 and is expected to be published in 2007.

2.3.4 Principles of Good Laboratory Practice 1999 (GLP)

In the UK, Statutory Instrument 1999/3106 contains both the current principles of GLP[9] and the procedures by which they will be implemented. The standard is maintained by the Organisation for Economic Co-operation and Development (OECD).[10] Revision of the OECD principles in 1998, which were then adopted by the EC as directives 99/11/EEC and 99/12/EEC late in 1998, led to an update of the UK regulations in 1999.[11]

The principles of GLP cover all the requirements of ISO 9001:2000 and ISO/IEC 17025:2005, but also contain additional needs. For example, a study plan is required in sufficient detail that the study can be recreated at any time in the future, and a study director is appointed as the single point of study control with responsibility for the overall conduct of the regulatory study and its final report. Due to the additional requirements of GLP, laboratories often ring-fence their GLP activities for ease of management.

In the UK, GLP Monitoring Authority (GLPMA) is the government body charged with enforcing GLP.[9] The GLPMA is made up of the Secretary of State for Health, the National Assembly for Wales, the Scottish Ministers and the Department of Health and Social Services for Northern Ireland. The work of the GLPMA is carried out by an executive agency of the Department of Health, the Medicines and Healthcare Products Regulatory Agency (MHRA).

2.3.5 Joint Code of Practice for Research

This code applies to contractors seeking funding from the FSA, DEFRA or the UK Devolved Administrations. The key areas covered by the code are the responsibilities of both the organisation and the project managers for the quality of work, competence of personnel, development of project plans in collaboration with the funding body and inclusion of risk assessments.[2] In addition, processes must be in place to assure the quality of research, all samples and experimental materials must be comprehensively labelled and traceable, the facilities and equipment must be appropriate for the measurements to be made, all procedures and methods must be documented and project leaders must regularly review the records of each scientist.[12]

2.4 Selection and Implementation of a Formal Management System

One of the most important points when implementing a management system is to match the standard to your work, not to change the work to match the standard.

The adoption of a management system should be a strategic decision for the organisation. It is always important to balance the advantages of gaining recognition for working according to a particular standard with the expense of implementation. The choice of system may be governed by a number of factors, including the nature of the business involved. For example ISO/IEC 17025:2005 may be the most appropriate standard for a testing laboratory, whereas ISO 9001:2000 may be more appropriate for a manufacturing business. The size of the business will also be a consideration, as a large laboratory may be so complex as to need more than one standard. Customer demand and the type of analysis to be carried out will also affect the decision. Many sectors are under legislative

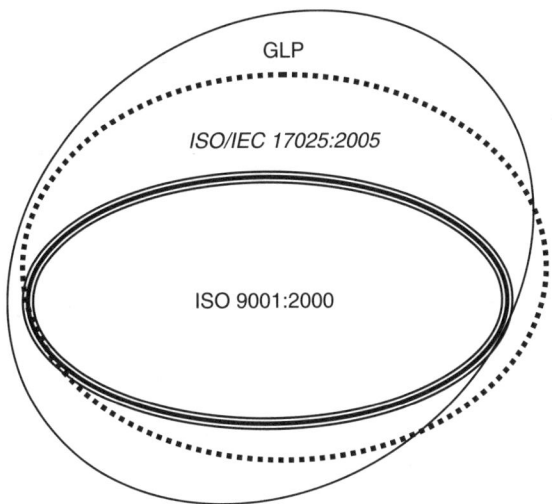

Figure 2.2 Schematic representation of the overlap in requirements for three international quality standards: ISO 9001:2000, ISO/IEC 17025:2005 and GLP.

control for particular tests and require specific accreditation before results and data produced will be accepted or testing can be carried out officially.

The remainder of this chapter covers many of the aspects that need to be considered when implementing a formal management system, although it is by no means an exhaustive list, and the information is not related directly to specific standards. Any discussion of the international standards will be limited to ISO 9001:2000, ISO/IEC 17025:2005 and GLP, unless otherwise stated. The three international standards all allude to the topics covered below, although the level of detail and documentation is different in each case (see Figure 2.2). The guidance is not intended to be a substitute for the official standards and guidelines, and the standards should be read directly if formal recognition is being considered.

2.4.1 The Management System

A management system is the formal structure set up to encompass all aspects of quality in the organisation. For the management system to be effective all the components (such as policies, systems, programmes, procedures and instructions) must be clearly documented and accessible so that everyone within the organisation is aware of the system and what is expected of them.

The international standards all promote the adoption of a process approach when developing, implementing and improving a business system, to enhance customer satisfaction by meeting customer requirements. In order to function effectively, most organisations must identify and manage numerous activities. A process is considered as a number of activities using managed resources in order to enable the transformation of inputs into outputs. A process approach is the application of a system of processes within an organisation, including their identification, interactions and their management, and is advantageous as it necessitates ongoing control over these processes. It may be helpful to follow a procedure such as 'Plan-Do-Check-Review', to implement and maintain a management system. 'Plan-Do-Check-Review' can be applied to all processes (see Figure 2.3).

In order to illustrate the principles and activities involved in implementing a management system more clearly, an example is given below. This shows how a medium-sized organisation involved in a diverse range of activities including testing, calibration, research and other services could implement a management system.

The company covers a wide range of activities, so it has to consider several international standards. The organisation as a whole is certified to ISO 9001:2000. The parts of the company concerned with analytical testing and calibration are accredited to ISO/IEC 17025:2005 and those parts of the organisation carrying out environmental fate and registration studies are registered to the OECD system of GLP.

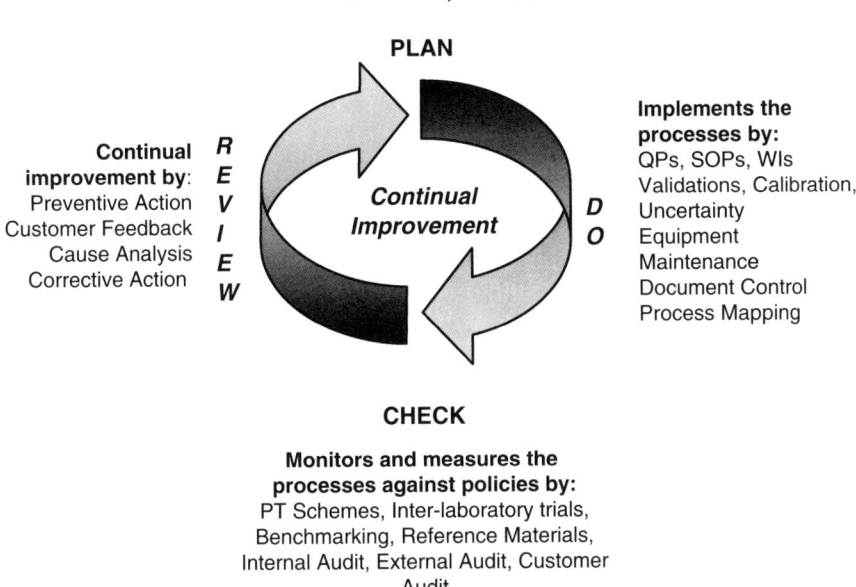

Figure 2.3 Schematic illustrating the 'Plan-Do-Check-Review' system of business management.

In order to implement a management system the organisation has performed a 'Plan-Do-Check-Review' process.

The system is described in a hierarchy of documents comprising at the top level a Quality Manual; there are also Quality Procedures (QPs), and then, where appropriate, Standard Operating Procedures (SOPs) and Work Instructions (WIs). The Manual, QPs and SOPs are all centrally controlled documents that are periodically reviewed and approved by authorised personnel before use; this level of control is widely adopted in routine testing laboratories. WIs are controlled at the local level; for example, an operating instruction for a specific piece of equipment. In a research laboratory WIs may be written for processes that are at a development stage. These WIs may progress to SOPs.

2.4.1.1 Quality Manual

The Quality Manual contains policies and summarises the principal elements of the management system. It describes how the laboratory meets the requirements of the management standards that apply. All the standards require an organisation to have a Quality Manual, but the information the manual should contain is not prescribed by the international standards described here; there is flexibility in the size and structure of the manual, which will vary depending on the complexity of the organisation. The Quality Manual used in our example explains the organisation's approach to certain issues, before describing the individual QPs. For example:

- Management responsibilities;
- Management of resources, including human resources, infrastructure and work environment;
- Planning and contracts, including negotiations with customers, customer requirements, purchasing, provision of service to customers and control of monitoring and measuring devices.

As well as the above, the manual will also hold organisational charts and descriptions of responsibilities for senior positions within the organisation.

The authorising person for the Quality Manual is the organisation's senior manager, for example the Chief Executive Officer.

2.4.1.2 *Quality Procedures (QPs)*

The principal elements of the system are set out in the QPs contained within the Quality Manual. These describe the implementation of significant company policies, which have an effect on the quality of service. The QPs describe what is done and who is responsible for the work. An example list of QPs contained within a Quality Manual is shown in Figure 2.4.

QPs are authorised by senior staff who are experts in the area of work, and who are responsible for reviewing them regularly. QPs are clearly labelled to aid in document control and should include the following: title; author; authoriser; issue number; date of issue; page number and total number of pages.

2.4.1.3 *Standard Operating Procedures (SOPs)*

These are detailed instructions, for analytical methods, calibration procedures and all GLP activities, that describe how the work is carried out. Of the four standards discussed, GLP is the most prescriptive about the categories to be covered specifically as SOPs. An SOP should be written in unambiguous language.[13] A typical SOP for a method may include the following sections:

- Title of the method;
- Scope of the method;
- Safety issues (related to chemicals, equipment and apparatus required);
- Reagents and materials required;
- Organisms used;
- Method procedure (including any calibration and validation);
- Quality control and quality assurance;
- Expression of results;
- Calculations;

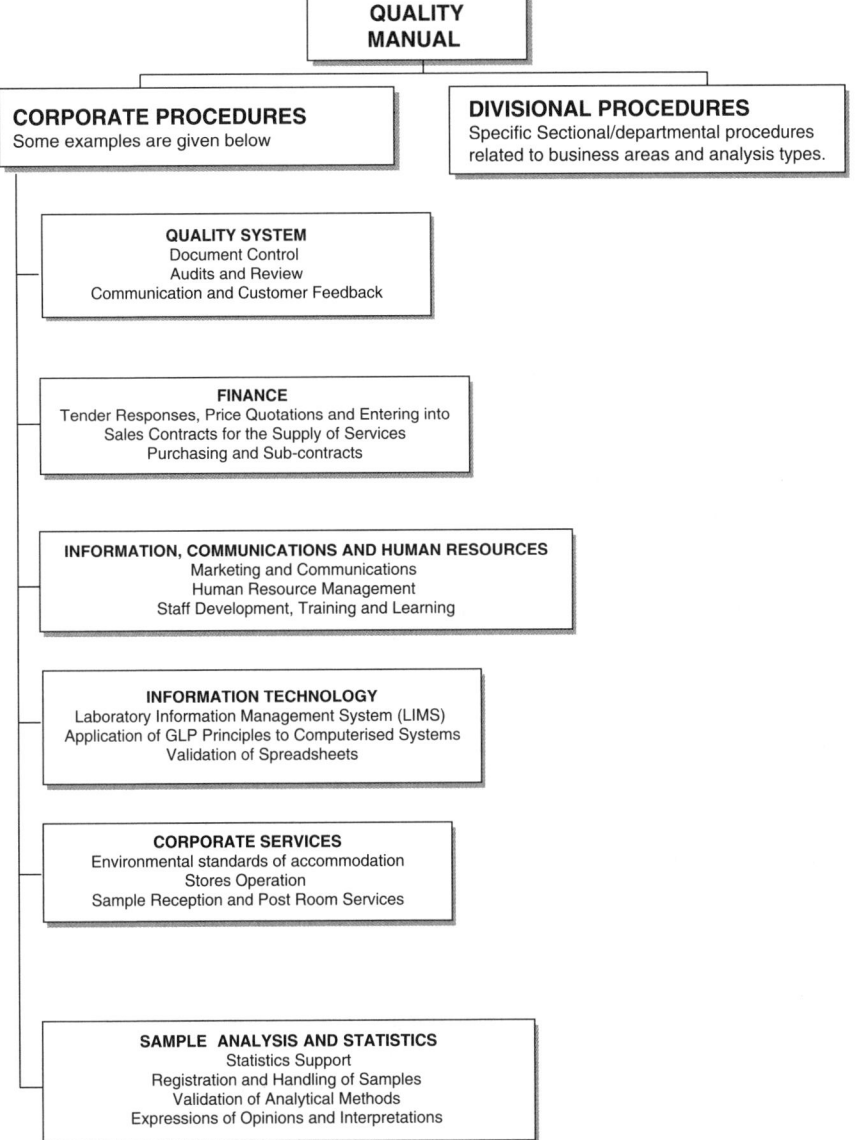

Figure 2.4 Listing and organisation of quality procedures, which may typically be contained within an organisation's Quality Manual.

- Performance data;
- Measurement uncertainty;
- Specific storage and disposal of standards and samples;
- Specific environmental control;
- Specific reporting requirements;
- References to other documentation such as QPs, WIs and other SOPs.

SOPs are authorised by senior staff and submitted to the SOP co-ordinator. They are also reviewed on a regular basis.

2.4.1.4 Locally Controlled Documentation

Locally controlled documentation is provided in the form of:

- Work instructions;
- Staff instructions;
- Training programme (local);
- Local team organisation (management structure).

These may comprise the following elements:

- Analytical method;
- Equipment calibration;
- Equipment operation.

These are all controlled documents subject to periodical review. Other locally stored documents are handbooks and manuals.

WIs are issued by senior staff, who are responsible for their maintenance and control.

2.4.2 Laboratory Environment

Accommodation needs to be controlled in a way that matches the analytical requirement, does not adversely affect the quality of data produced and maintains sample integrity. The infrastructure of the facility needs to be appropriately designed and of adequate capacity to contain the analysis required. Thought, therefore, needs to be given to all aspects of the process, including siting of equipment, containment issues, segregation, storage and archiving provision. Conditions which may affect the processes and the equipment being used need to be controlled, although these will be different depending on the process. Aspects that may need to be considered are air humidity, direction and rate of air flow, temperature, vibration, light intensity, power supply and space for operation of instruments. The space for an instrument to operate effectively may be greater than the machine' footprint. It is likely, especially where computers are involved, that equipment will require air circulation to prevent overheating.

2.4.2.1 Safety

Safety is also of the utmost importance within the laboratory, and many of the conditions considered above will also contribute to a safe working environment if controlled properly. A safe working environment is also necessary for staff to maintain quality. It is important, for example, to ensure flammable substances are stored correctly in order to minimise fire risk, or to ensure toxic substances are used in a fume cupboard to protect workers. The laboratory should be tidy so as to prevent accidents and to make it easier to trace reagents and samples through the analytical process. Bio-containment is also a very important quality and safety consideration, and guidance can be obtained from the UK Health and Safety Executive (HSE). It is vital to ensure the laboratory has the correct containment level for the samples being used, so that the environment and the workers are protected from biohazards. Effective containment should also prevent contamination of downstream processes which might otherwise affect the quality of the output.

2.4.2.2 Spatial Separation

Bio-containment is one specific type of physical separation, which is practised to prevent contamination. It is also important to consider the entire process and to physically separate areas where parts of the process may interfere with each other to the detriment of the quality of the result/product. Physical separation is one of the most widely cited means of controlling cross-contamination and yet is perhaps the most difficult to set up and adhere to. For example, in a molecular biology laboratory performing PCR, it is important to separate any pre-PCR sample preparation areas from the PCR reaction set-up process and, in turn, separate this from the post-PCR (PCR-positive) analysis areas of the laboratory. If space exists, there is also a very strong argument for using a fourth area, the template addition room, to ensure that PCR set-up and all associated reagents and consumables can be maintained in a DNA-free environment.

In a laboratory where physical separation exists a sample will 'flow through' the areas in a specific order. The 'flow' in such a laboratory would start at the pre-PCR, through to the PCR set-up/template addition areas and on to the PCR-positive area of the laboratory. Commercial, large-scale or routine services are recommended to use a 'one-way' progressive system for personnel involved in DNA extraction, PCR set-up and post-amplification analysis in order to enforce physical separation. In a forensic environment, for example, staff can be prevented from entering the pre-PCR or PCR set-up areas if they have already entered the PCR-positive area that day.

If possible the airflow through the laboratory should also be controlled, such that the direction of flow is the same as the sample flow; specifically, air should flow away from the pre-PCR and PCR set-up areas towards the PCR-positive area and not the other way around, thus preventing the passage of PCR-product back to the beginning of the process. This helps to prevent contamination by PCR products from previous reactions or other sources of DNA into the reactions being set up. In order to use such a system effectively, dedicated equipment, such as pipettes and racks, and consumables are required for each area, and the movement of DNA between sections by personnel may be limited by the use of area-restricted laboratory coats.

2.4.3 Equipment

2.4.3.1 Analytical Requirement

The primary requirement for all equipment is that it must be fit for purpose. Management standards require that instruments are suitable for their intended use and regulatory authorities and accreditation bodies are increasingly seeking and requiring evidence that equipment is fit for purpose. When choosing equipment for use in the laboratory, there are a number of factors that need to be taken into account in determining the practical and technical requirements. Experience gained by laboratory staff, or other laboratories in the field, in use of the equipment can be valuable when deciding whether the instrument is adequate for the task. The reliability should be researched and can often be gauged by talking to other laboratories already operating the instrument. It is important to check that the supplier will provide training and after-sales support and that the equipment will be compatible with existing equipment in the laboratory. It is also important to consider the computers that will be running the instruments. The role of the computer is increasing as technology advances, and is often relied upon for running the instrumentation correctly and analysing results. As such the hardware and software also need to be considered when checking the suitability of equipment, especially issues such as data security and version control. As instrumentation and related software quickly become outdated, it is necessary to check regularly that they still meet the analytical requirement.

2.4.3.2 'Ownership'

Once an item of equipment has been purchased and sited in the correct environment (see Section 2.4.2) then an 'owner' should be selected, and given responsibility for ensuring the equipment is maintained and serviced regularly. The 'owner' should be competent in the operation of the equipment through past experience and/or training by the manufacturer, and will be responsible for training other staff to a competent level. It is vital for all operators to be fully trained, have demonstrated competence and achieved authorised user status, so as not to damage the equipment or unwittingly affect the product output.

2.4.3.3 Log Books and Maintenance

All equipment, including computers, should have an associated log book containing the following information: a description of the equipment, manufacturer's serial number, date of receipt and installation, location, software and version installed and when updates have occurred. Equipment also needs to be maintained properly, which will include regular in-house maintenance as suggested by the manufacturer, annual servicing by the manufacturer or other approved organisation and calibration (see below) to ensure the metrological traceability of measurements. Maintenance of computers will include updating the software as new issues become available. A validated data set should be used to check comparability of calculated results between the 'new' and 'old' versions of the software. Similarly, when new software versions are released for the operation of instruments, comparability can be checked by running samples and conditions that have been verified using previous versions. Any information relating to servicing, calibration and maintenance, including software updates and checks, should be recorded in the equipment log book.

2.4.3.4 Calibration

Calibration is vital to underpin metrological traceability of results, and to ensure there are no unacceptable differences between a parameter's in-house measured value and the stated value. Equipment used for analytical measurements should be calibrated using calibration standards or certified reference materials (see Section 2.4.7) and used within their limitations of accuracy or capacity according to the manufacturer' instructions. Depending on the complexity of the equipment and the availability of appropriate reference materials and calibration standards, the calibration process may be performed in-house, by an external laboratory or by a visiting engineer. Lack of equipment calibration may affect the results of an analysis, as demonstrated in a PCR-based PT assessment.[14] Calibration data will help to demonstrate that the output of the process is known within a stated level of certainty. When working to GLP or ISO/IEC 17025:2005, it is essential to perform System Suitability Testing and QC checks prior to sample analysis. In a research laboratory, where possible, quality control materials should be used to check the correct functioning of equipment on a regular basis, and ideally within each run. The laboratory should have a schedule of calibration for all the equipment. Records detailing the regularity of checking and performance of equipment in such tests should be kept and are useful in demonstrating the reliability of analyses performed in the laboratory. Large laboratories may have a QP giving information about how the schedule of equipment maintenance is decided.

2.4.4 Reagents

2.4.4.1 Reagent Quality

As with equipment, care must be taken with all materials used in the laboratory to ensure they are fit for the intended purpose and do not contribute to unreliable results. All standards and reagents, including water, should be of the appropriate grade, purity or specification, used within the expiry date and obtained from a reputable source.

2.4.4.2 Storage Conditions

All materials should be labelled correctly and stored in an appropriate location, usually determined by the stability, lability and any hazards associated with the material. For example, solvents should be stored in a solvent safe to reduce fire risk, fluorescently labelled probes which are light sensitive should be stored in the dark and temperature-sensitive items should be stored at the correct temperature, for example a 0–8 °C refrigerator or −20 °C freezer.

2.4.4.3 Reagent Traceability

Materials bought from external suppliers should be traceable by lot numbers, batch numbers and data sheets.

Reagents made up within the laboratory, such as buffers, should be written in a laboratory notebook or a reagent preparation book and be traceable back to the original lot numbers of the constituents. Aliquots of suppliers' materials or reagents prepared in the laboratory should be clearly labelled with the analyst' initials, date prepared or aliquotted, expiry date and any special storage conditions.

2.4.4.4 Stability/Batch Comparability

Reagents which may be affected by freezing–thawing cycles, such as enzymes, reaction mixes, primers and probes, should be aliquotted on arrival, before storage in suitable sized aliquots such that repeated freezing–thawing is not necessary. It is also advisable to keep back one aliquot from a batch, which can then be used to check comparability of performance between batches, for example on receipt of a new batch. This ensures consistency across tests and will identify whether the new reagent batch is performing in the same way as the old batch, enabling any differences caused by new reagents to be quickly identified.

2.4.5 Analysts

2.4.5.1 Culture and Competence

To implement and maintain a management system that consistently produces a quality service, there needs to be a sustainable work force. All staff in an organisation should have defined roles and responsibilities. The staff should know the relevance and importance of their activities, as this will help to instil a culture of quality across the organisation. All management standards require proof of competence of the staff. Therefore it is essential that a training and development programme is in place and that its effectiveness can be demonstrated.

2.4.5.2 Training and Development

To ensure reliable analytical performance, all staff must be competent in the skills required for the job, or receive the training and support to become competent. Required skills include competence in communication (written and oral), report writing, statistical analysis and use of IT, as well as the technical skills required to carry out the laboratory analysis. Organisations will, therefore, need to be committed to developing their work force. An effective route to ensuring effective staff development is to instigate at least an annual review to identify training needs and to formulate a training plan, with a mid-term check to note progress.

It is also vital that, having trained the work force, records of their skills and competencies are maintained to support the results being produced for customers. Each member of staff should, therefore, have a training record that is filled in during and on completion of training, which is signed by the trainee and the trainer to say that the required level of competence has been achieved. A training programme should be defined for each team, covering the knowledge, skills, techniques and methodology necessary to carry out the work. The programme will normally differentiate between the induction training for new staff and the ongoing training and development of all staff (see Figure 2.5). Where possible, objective criteria should be specified to enable judgement of acquired skills to be made.

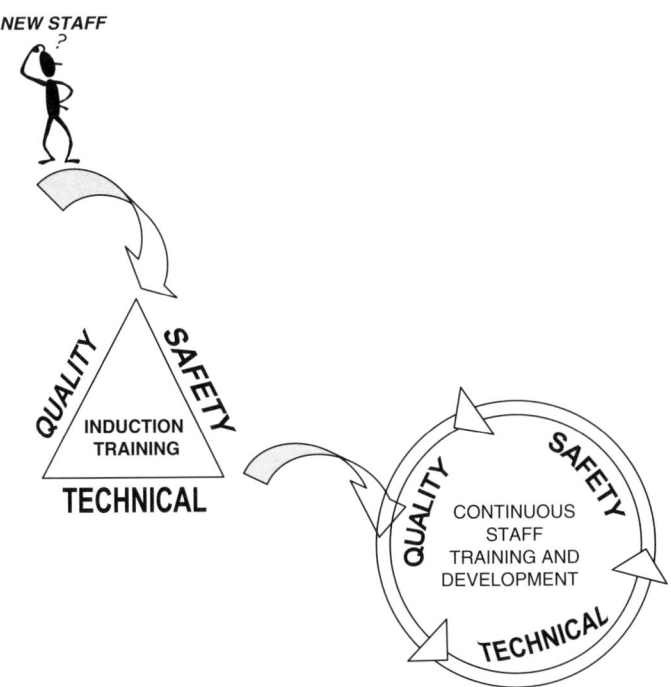

Figure 2.5 Schematic overview of a typical training programme, including the initial induction of new staff and the continual development of all staff.

2.4.6 Methods

2.4.6.1 Fitness for Purpose

Regardless of the management standards adhered to or the type of work being carried out, it is important to use methods that are fit for the analytical purpose. This applies to all procedures, from those used in the initial sampling, through handling, transport, storage, preparation, analysis and reporting. In ISO/IEC 17025:2005 there is a definite hierarchy of methods that should be used, starting with international standards at the top, followed by published methods and then laboratory developed methods. For a method to be deemed fit for purpose it is necessary to perform method validation (see Chapter 3).

2.4.6.2 Documentation

To demonstrate the quality of results, it is also important to be able to trace the methods, from the original validation data to any deviations from the methods that have been introduced. This is normally achieved by documenting all experiments in a laboratory notebook. Documentation usually requires the detailing of the method used, including a published reference and any deviations from the method, with the reasons for the deviations. It is important to recognise that if deviation from the original method occurs, the method will no longer be validated and extra work will be required to establish that the results obtained are fit for purpose.

2.4.6.3 Metrological Traceability

It is desirable to be able to trace the results from experimental processes back through an unbroken chain of comparison, each with a stated uncertainty, to national or international standards. Traceability can be achieved through the use of certified reference materials. However, the lack of availability of suitable reference materials precludes this approach for most molecular analyses (for more information see Section 2.4.7.1). When this approach is not possible metrological traceability may be achieved by using a validated standard method.[15,16]

2.4.6.4 Independent Quality Assessment

It is becoming more widely recognised that there is a need for laboratories performing nucleic acid measurements to utilise independent means of assessing their performance and assuring quality.[17] Analytical performance may be demonstrated by taking part in Proficiency Testing (PT) or External Quality Assessment (EQA) schemes. Such schemes generally involve distribution of a test material, known to be stable and homogeneous, to participating laboratories to be analysed for one or more target analytes, using normal in-house methods, analysts and equipment. Participants return their results for evaluation by an agreed deadline, and comparative performance of all participants is then documented in a report given to all participants. Usually each participant is given a unique identification code, allowing comparison of individual results with peer laboratories whilst maintaining confidentiality. Performance may also be measured by comparison of the observed result with the true value, if sufficiently characterised samples were provided in the scheme. Participation in independent assessment schemes can produce a general improvement in analytical technique through the identification of errors, and may also be used for analyst training or evaluation of different analytical approaches by participating organisations. Participating in PT schemes, if they are available, is a requirement of ISO/IEC 17025:2005.

2.4.6.5 Method Validation

If a method is being used to report analytical results to customers it is important to understand the scope of the method and to be sure that the method fulfils the specified requirements. The process of ensuring these requirements are met is known as method validation and is a requirement for testing or calibration laboratories working to ISO/IEC 17025:2005.

The validation process defines a range of performance characteristics within which it has been confirmed that the method can yield acceptable results. The use of a valid method is, therefore, limited by its application. Validation does not just apply to the instrumentation, but to any computer software associated with the process. The actual procedures for method validation may vary from sector to sector. It is advisable to follow any specific sectoral guidance if available, to ensure comparability with peer laboratories. More detailed information on method validation is provided in Chapter 3.

2.4.6.6 Experimental Design

Use of effective experimental design is key to ensuring confidence in results, enabling efficient generation of sufficient data for meaningful statistical analysis. Effective design enables reliable conclusions to be drawn from experimental results, with concomitant savings in time and money. In addition to ensuring sufficient data are produced for routine analyses, design principles may usefully also be applied to method optimisation. Performance of a carefully selected set of experiments, in which all relevant factors can be varied simultaneously, together with appropriate statistical analysis, allows for any interaction between factors to be fully investigated and optimum conditions determined. The size and complexity of an experiment may necessitate the use of dedicated experimental design software.

Although there are several criteria that can be considered when designing the experiment, the following three should always be considered.

- *Controls*. The effectiveness of a technique is often assessed by way of comparisons. The nature and source of samples of interest versus controls should, therefore, be carefully planned.

- *Randomisation.* The chief purpose of randomisation is to provide an unbiased estimate of sample values, and hence randomisation should be implemented whenever practicable. In practice, randomisation may be technically more demanding, as extra care and time may be required in experimental set-up if samples are fully randomised within an experiment, as replicates of the same sample type will not necessarily be placed next to each other. It may also be necessary to conduct the experiment 'blindly', with the analyst unaware of sample identity, so that any possible bias due to user interpretation is reduced.
- *Replicates.* Replication has a two-fold function within experiments. Firstly, replication enables statistical analysis to be performed on a set of results, as a single reaction or duplication provides insufficient data for statistical analysis. Secondly, replication increases the power with which comparisons of interest are made. In general, the higher the replication factor for a comparison of interest, the more confidence can be attributed to the decision or interpretation based on that comparison.

As a rule of thumb, a replication factor of at least six is recommended for any one comparison of interest. Statistically, six replicates will provide a reasonable representation of the mean and variance of the original population from which the six replicates were drawn.

2.4.6.7 Measurement Uncertainty

Estimation of measurement uncertainty is a requirement of ISO/IEC 17025:2005, but not necessarily ISO 9001:2000, and may also be requested by customers. Rather than casting doubt on the result of an analysis, provision of an uncertainty value allows the possible variability from the 'true' result to be assessed and the data to be interpreted accordingly. More detail on the derivation and use of measurement uncertainty is given in Chapter 3. In addition, guides are available which may help in the understanding and assessment of uncertainty values.[18,19]

2.4.7 Quality Control

Quality Control (QC) procedures provide routes to checking the validity of processes on a day-to-day basis, ensuring that results produced are within known probability limits for accuracy and precision. A range of sample types, including negative controls, QC samples and blind samples (see Section 2.4.7.2), can be employed and run alongside the routine test samples, in order to check the quality of the analytical process. Monitoring analytical performance through the use of blind samples, duplicate analysis and control charting are also effective QC processes.

2.4.7.1 Reference Materials

A reference material (RM) is defined in ISO Guide 35[20] as:

'Material sufficiently homogeneous and stable with respect to one or more specified properties, which has been established to be fit for its intended use in a measurement process.'

The purpose of an RM is primarily to facilitate metrologically traceable measurement results, rather than acting as a positive control for an analytical process and, as such, forms part of the QA system. Once the property value(s) of a particular RM have been established by measurement, they are, in effect, 'stored' by the reference material up until its expiry date. RMs are therefore valuable resources used to underpin consistency and comparability of analyses.

Certified reference materials (CRMs) are also sometimes referred to as, and are defined as: 'A reference material characterised by a metrologically valid procedure for one or more specified properties, accompanied by a certificate that states the value of the specified property, its associated uncertainty and a statement of metrological traceability.' The reader is referred to 'Quality in the analytical chemistry laboratory' for a more in-depth discussion on RMs.[21]

The complex molecular composition of biological materials, coupled with the inherent heterogeneity, potential lack of stability and difficulty of characterisation, makes fulfilment of the stringent requirements of an RM especially challenging.

Despite the increased reliance on nucleic-acid-based analyses in a variety of sectors, the availability of high-quality reference standards is very limited. The paucity of such materials is attributable in part to the pace of technical development in the field, partly to the considerable cost in development and production of certified reference materials, and also to the complexity and inherent instability of biological materials. Thus there are many measurements in analytical molecular biology for which an appropriate reference material does not exist at this time.

However, there is a role for commercial standards for some applications. Typical commercial standards employed in analytical molecular biology include DNA and protein molecular size markers and allelic ladders for genotyping applications, high purity DNA/RNA samples of known origin and numerous positive control components supplied with a vast array of detection or analysis kits. The range of commercial standards is increasing in response to the greater demand for controlled and traceable measurement results. However, reliance on commercial standards may be problematic, in terms of the homogeneity, stability and reproducibility of potentially less-well-characterised bulk commercial products. Producers of RMs are encouraged to meet the requirements of ISO Guide 34.[22]

2.4.7.2 In-house Quality Control Materials

Materials frequently used as QC samples include: spiked matrix samples, replicates and positive and negative control samples (these may include negative samples analysed during the measurement process or negative samples that have been carried though the entire analytical procedure, for example from sample extraction through to measurement). Duplicate and replicate samples allow precision of measurements to be assessed, negative controls inform about potential contamination of the analysis and known positive samples can indicate false negative measurements caused by inhibition, equipment failure or analyst error. In-house materials have not necessarily undergone the rigorous tests, characterisations and documentation required of a reference material but provide a useful alternative to high-level RMs, allowing performance control materials to be tailored to a particular analyte or process. Such materials should be stable and homogenous and, if calibrated against RMs, in-house standards can also reduce the cost associated with the use of RMs by extending the usage of a single RM supply. However, for continuity a reasonable amount of the in-house standard needs to be produced in a single batch.

The use of QC samples provides assurance that the method is performing acceptably. QC samples should be stored, handled and analysed in parallel with the test samples, and the level of analyte in QC samples should be in the same range as that in the samples under test. QC samples may range from in-house previously characterised samples to commercially available, characterised samples of established and certified measurement value. As a guide, QC samples should make up approximately 10–20% of the samples analysed, although this could be decreased for robust high-throughput types of analyses or increased to 50% when carrying out complex *ad hoc* forensic analyses. Results from QC samples should be checked prior to assessing data from unknown samples. Any deviation from expected results would require corrective action and the co-analysed batch of samples would require re-analysis.

To provide realistic samples for QC, a variety of biological matrices (such as soil, water, foodstuffs and biological fluids) can be spiked with target cells or DNA. Admixtures of matrices can also be used to mimic real situations. Although such materials offer the flexibility to match the controls to the actual samples under analysis, disadvantages of this approach include the difficulty of reliably preparing homogenous samples and batch-to-batch variation. During the process of spiking, where analytes are added directly to the biological matrix, the targets may not react predictably in their new environment. For example, a considerable percentage of cells may lyse and start to become degraded, as it is difficult to simulate actual matrix conditions such as pH, temperature and oxygen content. In addition, spiked samples may have different physical properties from those of real samples where the cells or nucleic acids may have been exposed to the matrix for a much greater period of time.

2.4.7.3 *Performance Control*

Control charts are routinely used to record data obtained from control analyses, and are an effective monitoring tool. For routine analyses, one or more indicators of analytical performance are usually plotted against time or number of measurements in a graphical display. Often the average values and upper and lower acceptable limits, representing a given variation from average (commonly 3 × standard deviation), are included in the plot. Such charting can rapidly identify deviations from normal or expected analytical performance, highlighting differences between analysts, degradation issues or batch-to-batch variation of reagents and equipment drift or failure. Non-random distribution of measurements may also become apparent through control charts, and may indicate regular seasonal or other variations affecting the results of the measurement process.[23]

2.4.7.4 Contamination Control

Contamination can be a major source of poor quality data in a molecular biology laboratory, and it is important to design experiments with sufficient negative controls to identify if contamination has occurred. Sensible laboratory practice and good housekeeping can largely avoid the problem of contamination. For example, protective clothing dedicated to each area should be worn by all occupants of the laboratory, single-use aliquots of reagents should be routinely prepared and used, samples should remain sealed at all times when not in use, the production of aerosols or fine, particulate biological material should be avoided or manipulations carried out in a suitably contained space and all equipment, utensils, materials and benches should be regularly cleaned or sterilised as appropriate.

The layout of the laboratory may require areas designated to a certain task in order to avoid cross-contamination (as described in Section 2.4.2.2). Dedicated equipment and protective clothing should be freely available in each location, thereby minimising contamination by the movement of materials between areas. Other activities, such as RNA extraction and analysis may also benefit from spatial separation and dedicated equipment. In laboratories carrying out analysis of human DNA, samples from analysts performing the work should be characterised to enable contamination by laboratory staff to be identified if it occurs. In addition, if contamination is a persistent problem, investigation of all reagents may be prudent, as it has, for example, been documented that some commercial preparations of *Taq* polymerase have been insufficiently purified during manufacture, resulting in persisting contamination with bacterial genomic DNA.[24]

2.4.8 Samples

A number of factors need to be considered in handling samples prior to analysis, as changes in sample integrity, poor labelling and inappropriate sampling and preparation procedures may significant affect the results of the analysis.

2.4.8.1 Chain of Custody

To ensure traceability to source, a sample should be entered into a registration system immediately on arrival at the analytical laboratory. Registers for samples submitted for analysis should contain information such as the name of the customer, customer reference number, unique identifier assigned by registration team, sample description and any notable features, date received, who received by, agreed and actual report dates, sample disposal date and route and any special instructions or requirements.

Once the sample has been entered into the registration system, this unique identifying number should be used to follow the sample through the analysis process to the reporting of the results. This number should, therefore, be used to reference the sample in laboratory notebooks, on storage tubes, during testing, during data analysis and on the final report to the customer.

Laboratories that have a very high throughput of samples may choose to use a Laboratory Information Management System (LIMS), instead of a paper register, to keep track of the samples. The LIMS is used in laboratories to record all relevant sample and batch information; it enables efficient reporting of results to customers and simplifies checking sample information without having to look through manual registers. A sample is entered into the LIMS and assigned a unique identifier; a bar code can then be printed and attached to the sample. This bar code is then used to trace the sample through the system. Whenever a sample is moved or an analysis carried out, this information is entered into the system. It is therefore possible to gain a full history of the analyses carried out and the passage of the sample through the entire process.

2.4.8.2 *Sampling and Preparation*

The largest error in any analysis often stems from the sampling process. It is essential that the material selected for analysis is truly representative of the entire sample with respect to both matrix and analyte.[25] The effect of the initial sampling on the result may be significant, but the process may not be in the control of the analyst. However, every effort should be taken, during contract review, to give as much specification as possible to the customer regarding the initial sampling and how it can affect the analysis. Use of appropriate sampling procedures ensures the samples received by the laboratory are suitable for the analysis and reporting level required. The sampling employed needs to be appropriate to the analyte and the matrix, with sample size and analyte level being considered.

Both the extraction process and measurement process can be severely affected by sampling bias. To overcome this it may be necessary to ensure sample homogeneity prior to sampling. For a routine testing laboratory, demonstration of homogeneity will be part of method validation.[26] A method is normally stated as being fit for purpose for a specified sample size.

2.4.8.3 *Storage*

The storage of a sample before submission for analysis is normally out of the control of the analyst but, once submitted to the laboratory, every effort should be made to keep the sample in a suitable and stable environment. There are often specified requirements for storage of forensic, microbiological and radiochemical samples. The storage locations for these samples are approved by the appropriate contract controller. The testing laboratory should clarify with the customer, where possible, the favoured storage conditions for the samples to be analysed. In all cases the laboratory should attempt to store the samples in such a way as to avoid degradation and contamination.

When performing DNA extraction, the temperature at which the sample has been stored can greatly affect the sample quality and extraction yield. Freezing samples can prevent further growth of contaminating micro-organisms, whilst naturally occurring autolysis and DNA degradation by endogenous and exogenous enzymes can also be abated by reduced temperatures. Samples requiring long-term storage should be placed in an environment in which integrity is known to be maintained and inherent enzyme activity is arrested. Any freezing procedure should be undertaken as soon as possible after the arrival of the sample. Excessive cycles of sample freeze/thawing should be avoided, however, as this may induce cell or DNA breakdown.

The retention time of samples will also depend on the sample type and should be discussed with the customer. Sample disposal by the analytical laboratory can be classified as sample destruction, or returning the sample to the customer. In both these cases, the mode of disposal, the date and analyst carrying out the disposal should be recorded in the sample register to ensure complete traceability of the sample.

2.4.9 Recording and Reporting

Recording is a crucial part of any analysis; the laboratory notebook is one of the main places information and data are recorded. For a particular analysis the following information may be noted: Samples analysed, reagent batch number, equipment, methodology and observations, raw data collection and data analysis and interpretation.

2.4.9.1 Electronic Data and Automated Analysis

A vast amount of data may be generated during analysis and stored electronically. To maintain traceability, a cross-referencing system must be in operation, to link the data stored electronically with the data stored in the notebook and the customer report. For data files to be easily found and identified a relevant file structure should be implemented, including analyst's initials and date.

The raw data file and any subsequent data manipulations should be stored electronically. A back-up copy of all data files should be made and stored in a secure location. These protection mechanisms need to be in place to protect the data from being lost if a computer failure occurs. Data files should be protected from unauthorised change. Any changes made need to be visible and evidence of the changes and reasons for them need to be recorded. Password protecting documents can prevent unauthorised access. Further information on long-term data storage is discussed in Section 2.4.10.

Equipment-associated software may perform some form of data manipulation before it is accessible to the analyst, in which case the initial raw data are not available to the analyst. Real time PCR data, for example, may be presented with a background correction already applied. It is, therefore, important to critically assess the presented data and have an understanding of the manipulations that are made and how these may differ from data manipulation processes in other similar instruments, potentially affecting the comparability of results.

Several software packages may be used to analyse fully the data produced from an experimental run. This presents the opportunity for the data to be inadvertently altered during the transfer from one software package to the other. It is, therefore, important that all transcription of information, whether manual or electronic, be checked carefully to ensure the data are transferred intact and unaltered.

2.4.9.2 Reporting

The analytical report should clearly address the customer' request and should be accurate, concise, unambiguous and include all the information required by the customer for full interpretation. Opinions and interpretations quoted in the analytical report are covered specifically in ISO/IEC 17025:2005. However, neither the opinions nor the interpretations given in a report are in themselves accredited. The laboratory which provides the opinions and interpretations is accredited to do so, on the basis of a set of criteria. The laboratory will have a process of selecting individuals who through experience and qualifications are able to provide opinions and interpretations for specific tests. It is this process that is accredited.

Reports will normally be reviewed by another individual familiar with the analyses conducted and the type of results being reported and then authorised by a senior staff member before being despatched to the customer. In order to trace all reports produced by the organisation, unique identifying numbers can be given to each report submitted by the organisation to a customer. This number is then entered into the sample register to complete the entry for a particular sample or batch of samples, thus completing the cycle of traceability.

2.4.10 Archiving

Samples and data may both need to be stored for a period of time after the initial analysis has been completed and reported. This period may be determined by the organisation to meet customer requirements (ISO 9001:2000, ISO/IEC 17025:2005).

In all cases, if samples, data or paperwork are archived, a filing system needs to be in place, such that locating the item and, therefore, retrieval, is possible.

The environment of the storage facility may need to be controlled to ensure that no degradation occurs, and may include control of temperature, humidity and light. The requirements of GLP for archiving material are much more stringent than ISO 9001:2000 and ISO/IEC 17025:2005. However, discussion of these requirements is beyond the scope of this chapter and the reader is referred to 'Good Laboratory Practice, OECD Principles and Guidance for Compliance Monitoring'.[11]

2.4.10.1 Electronic Data

Copying (backing up) data onto tapes, or other such media, will ensure that there is a copy of the data, even if the computer fails. Care needs to be taken that a suitable media type is used for the storage and that the data will be retrievable in several years' time. A 5-inch floppy disk, for example, can now no longer be read by most computers and yet may have been used to store data 6–7 years ago. The manufacturer' warranty length will also affect the ability to retrieve the data in the future. There is no point storing data for 20 years if the manufacturer' warranty only guarantees retrieval for 10 years. It is important to note that different media types and different makes of the same media type may all have different life spans.

In order to increase the likelihood of retrieval, at least two copies of the data should be stored. It is sensible that these copies be stored in physically separate places, in a fire-resistant location such as a fire safe. Ideally one copy should be stored off site, but if this is not possible, they should, at least, be stored in separate parts of a building. This decreases the risk of all copies being destroyed together.

Consideration should be given to the software program being used to store the data, when archiving data. Programs are likely to change considerably over a 10-year period. Thought should, therefore, be given to the formatting of the stored data. For example, it may be wise to store the file in a tab-delimited format, which should be readable by numerous programs, rather than as a format specific to an individual program. It is also wise to check archived data periodically for readability on new software issues. For the very same reason, thought should also be given before completely destroying older software versions.

2.5 Summary

The information presented here is intended to outline the benefits of introducing and maintaining some form of management system to underpin the production of reliable results. For many laboratories obtaining accreditation to national or international standards will not be necessary. However, an understanding of the basic principles of Quality Assurance as applied to molecular analysis is central to improving analytical consistency, enabling effective research and analysis to be performed. Initial time spent in ensuring analytical quality will be more than outweighed by subsequent savings in time and cost by avoiding repeat analyses and uninterpretable results.

Acknowledgements

The author would like to thank Elizabeth Prichard for her invaluable contribution to the preparation of this chapter.

References

1. http://www.rsc.org/pdf/emeetings/aamg_outsourcing/lyne.pdf (accessed July 2007).
2. http://www.defra.gov.uk/science/how/documents/JCoPRGv02.04.pdf (accessed July 2007).
3. ISO/IEC 17000:2004, *Conformity Assessment – Vocabulary and General Principles*, International Organization for Standardization, Geneva, 2004.
4. ISO/IEC 9001:2000, *Quality Management Systems – Requirements*, International Organization for Standardization, Geneva, 2000.
5. http://emea.bsi-global.com/Quality/QMSregistration/BSIRoutetoReg.xalter/ (accessed March 2007).
6. ISO/IEC 17025:2005, *General Requirements for the Competence of Testing and Calibration Laboratories*, International Organization for Standardization, Geneva, 2005.
7. http://www.ukas.com/information_centre/accreditation_category_forms.asp/ (accessed July 2007).
8. ISO 15189:2003, *Medical Laboratories – Particular Requirements for Quality and Competence*, International Organization for Standardization, Geneva, 2003.
9. http://www.mhra.gov.uk/home/groups/comms-ic/documents/publication/con007566.pdf (accessed July 2007).
10. http://eescopinions.eesc.europa.eu/eescopiniondocument.aspx?language=en&docnr=75&year=2003 (accessed July 2007).
11. OECD, *Good Laboratory Practice, OECD Principles and Guidance for Compliance Monitoring*, 1st edition, 2005, ISBN 9264012826.
12. A. Fazackerley, *The Scientist*, 2004, **40**, 1.
13. ISO/IEC 78-2:1999 *Chemistry – Layouts for Standards – Part 2:Methods of Chemical Analysis*, International Organization for Standardization, Geneva, 1999.
14. G. C. Saunders, J. Dukes, H. C. Parkes and J. H. Cornett, *Clin. Chem. (Washington, DC, United States)*, 2001, **47**, 47.
15. http://www.vam.org.uk/publications/publications_item.asp?intPublicationID=839 (accessed July 2007).
16. http://www.measurementuncertainty.org/mu/EC_Trace_2003_print.pdf (accessed July 2007).
17. P. E. Garret, *J. Clin. Virol.*, 2001, **20**, 15.
18. http://www.measurementuncertainty.org/mu/guide/index.html (accessed July 2007).

19. ISO/IEC Guide 98:1995 *Guide to the Expression of Uncertainty in Measurement (GUM)*, International Organization for Standardization, Geneva, 1995.
20. ISO/IEC Guide 35:2006 *Reference Materials – General and Statistical Principles for Certification*, International Organization for Standardization, Geneva, 2006.
21. V. Barwick and E. F. Prichard, *Quality in the Analytical Chemistry Laboratory*, Wiley, Chichester, UK, 2007.
22. ISO/IEC Guide 34:2000 *General Requirements for the Competence of Reference Material Producers*, International Organization for Standardization, Geneva, 2000.
23. J. N. Miller and J. C. Miller, *Statistics and Chemometrics for Analytical Chemistry*, Prentice Hall, 5th edn, 2005, ISBN-13: 978-0131-291928.
24. C. E. Corless, M. Guiver, R. Borrow, V. Edwards-Jones, E. B. Kaczmarski and A. J. Fox, *J. Clin. Microbiol.*, 2000, **38**, 1747.
25. RSC, *General Principles of Good Sampling Practice*, The Royal Society of Chemistry, Cambridge, 1995, ISSN 0 85404 412 4.
26. T. Fearn and M. Thompson, *The Analyst*, 2001, **126**, 1414.

CHAPTER 3
An Introduction to Method Validation

SALLY L. HOPKINS AND VICKI BARWICK

LGC, Queens Road, Teddington, TW11 0LY

3.1 Introduction

Method validation is the practical process of determining the suitability of a method for providing analytical data that is fit for the intended purpose. Method validation is defined in ISO/IEC 17025:2005[1] as:

> 'Confirmation by examination and the provision of objective evidence that the particular requirements for a specific intended use are fulfilled.'

For any method to produce meaningful and reliable data, some performance checks should be made before the method is applied to real samples. The validation process typically entails firstly understanding the reason the measurements are being made and the performance of the method that is required to produce data that are fit for that intended purpose. Secondly experiments are planned and performed to evaluate the performance of the method. The observed performance is then compared with the required performance of the method, and relevant/specified criteria are used to determine whether the performance is adequate. The actual level of assessment and validation that is undertaken will depend on the intended use of the method and the importance of the data produced.

It is therefore important to understand the required scope of the method and to be sure that the method fulfils the specified requirements. In practice, the scope of a method refers to the type of samples that are analysed (both the nature and level of target that is present and the sample matrix). The specified requirements indicate the performance that is required to deliver results that are fit for the intended purpose, which, for example, may necessitate that the method can detect a minimum amount of target, or give measurements that are very close to the real value of the target present in the sample. Validation is specific to the particular combinations of sample matrix and analyte covered by the method scope, and the equipment used in the validation process, and therefore defines acceptable performance characteristics within those analytical limitations. Extrapolation to other sample types, equipment or laboratories will necessitate re-validation or verification that the method is still suitable under the changed conditions.

In validating a method it is assumed that analysts are trained and competent, that the laboratory environment is appropriate, and that any instrumentation and software used is suitable for the task, well maintained and calibrated correctly. These requirements may necessitate a series of pre-validation checks and calibrations, as evidence will need to be obtained to demonstrate that this is the case if not already available (see Chapter 2). The formal process of demonstrating that an item of equipment is suitable for a particular application is referred to as 'equipment qualification'.

In chemical analysis the process of validating a method is well established. By contrast method validation in biological analyses is often more challenging because of the complexity of the analytes, and the paucity of reference materials, standards, established methods and examples of the process. This chapter is intended to introduce the processes and terminology typically used in method validation, and to indicate their application to biological analyses. Some of the key issues associated with method validation are highlighted, although specific sectoral guidance should also be consulted if available. Several publications are also available giving much more detail on all aspects of validation.[2-5]

3.1.1 Why and When is Method Validation Necessary?

Method validation may be necessary for a number of reasons. Ethically method validation is important, as a 'customer' employs a laboratory to carry out analysis on its behalf. The customer in reality may be a funding body for academic research or a laboratory colleague, and essentially refers to any end users of the results of the experiment. The laboratory or researcher should, therefore, ensure that the method is fit for the intended use through validation. Without validation there are no guarantees that a method will produce reliable or meaningful data, undermining the purpose of the analysis.

Legally method validation is a requirement for work carried out under GLP[6] (Good Laboratory Practice) guidelines and it is also a requirement for any methods accredited under ISO/IEC 17025. ISO/IEC 9001[7] also requires validation of any processes for production and service provision where the resulting output cannot be verified by subsequent monitoring or measurement.

Commercially it makes sense to have assurance that a method will work as expected and the quality of the output is suitably controlled. Validation helps to provide this assurance, and prevents wasting money on measurements made using unsuitable methods which then have to be repeated. Measurement of method performance parameters during validation can also help to identify the critical stages of the method which require tight control. Validation, therefore, helps in the design and implementation of suitable quality control procedures to ensure the long-term reliability of test results.

Method validation normally expands on information acquired during the method development stage, when the method is assessed to ensure it is suitable for the analytical task at hand. Once this has been established, method development gives way to more formal validation studies. Analysts should always check the level of validation previously carried out on published methods and undertake additional studies if the existing data are insufficient to demonstrate fitness for purpose. However, even if the validation of published or standard methods is considered sufficient, individual laboratories should always carry out some evaluation of method performance in-house prior to use to ensure that the performance is adequate. This process is sometimes referred to as verification.

It is not only new methods that require validation. Reinstating a method after a period of non-use will require validation checks to be carried out to demonstrate that the performance is still acceptable. Extending the scope of the method to include different sample matrix types or different analyte levels, for example, will require the performance to be checked using the new sample types, as will any other parameter changes such as the laboratory environment, analyst or instrument. The amount of method validation carried out for a particular method will depend on the circumstances and requirements for the particular analytical problem, and may be influenced by four main considerations, as follows.

3.1.1.1 Criticality of the Data

The intended use of the data from an analysis determines the criticality, that is, the level of importance that the result is correct and reliable. For example, results from samples tested from a scene of crime for identification purposes or analysis of biopharmaceutical products to determine safety would both be considered critical. However, all data are critical to a degree, as there is always a reason for performing the analysis, so use of invalid or untried methods is not an acceptable option.

3.1.1.2 Uniqueness of the Sample

A sample might be considered unique in terms of time (representing a unique moment in the progression of an infection), quantity (single hair at scene of crime) or difficulty in obtaining (biopsy material).

3.1.1.3 Robustness of the Technique

Well-established and easy-to-use methods will generally require less validation than more complex and technically demanding methods with many critical points. The less influence that experimental variability has on the outcome of a method, and the greater the robustness, the less validation will be required.

3.1.1.4 Expected Level of Utilisation of the Technique

Analysis performed on a small number of samples as a 'one-off' test may not require extensive validation, as factors such as long-term reliability and inter-operator variability will not be an issue.

However, even if neither the method nor the sample is judged critical, some level of evaluation of method performance is required. Generally professional judgement and experience are used to balance time and resource constraints against validation requirements. The fundamental aim of getting the analysis right first time, and every time, can be facilitated by using appropriate and validated methodologies.

3.2 Planning the Validation Process

The first step in any validation should be to consider the problem presented by the end user of the data. Why is the analysis being carried out, what is expected from the analysis? Once you know the problem that the data from your analytical measurements are intended to solve, you can determine what is expected from the method. This enables a suitable method to be chosen (considering constraints such as time, laboratory resources, sample size and complexity) and the method performance characteristics which are relevant to the work to be determined. The method is scoped by defining the samples to be analysed, the target and levels that are expected and the confidence that is required in the results. Any additional constraints are also considered, and then a detailed validation plan can be made.

Often several parameters can be examined in one set of experiments, so the order in which experiments are carried out must be logically planned. Experimental design software can facilitate effective planning, and is especially useful for complex experimental designs. Specialised software packages, such as the Modde series of software available from Umetrics, have been designed in order to produce the most efficient experiment based upon a number of user defined criteria. mVAL, software to facilitate method validation, is also available.[4] However the user must be familiar with the specialist terminology of both the software package and statistics in order to use such programs effectively.

Once the plan is finalised, experiments are performed to evaluate the method performance parameters, and the data produced are used to determine whether the method is fit for its intended purpose.

3.3 Method Performance Parameters

There are many performance characteristics that can potentially be investigated for a particular method, some of which are listed in Table 3.1, and described in more detail in the following sections. Various performance parameters are

Table 3.1 Range of performance characteristics that may be evaluated in method validation.

- Precision
  - Repeatability
  - Reproducibility
- Bias
- Recovery
- Accuracy (precision + trueness)
- Ruggedness and robustness
- Selectivity (specificity)
- Limit of detection (LoD) or sensitivity
- Limit of quantification (LoQ)
- Linearity
- Working range

---

important depending on the type of measurements being made and the reasons for making the measurements, so choosing the characteristics to be investigated is a crucial part of the validation process. Accuracy (comprising both precision and bias) may be important for calculating absolute values of properties or analytes, for example, whereas precision is more significant in comparative studies. Random errors in the method are reflected in the precision of the results, whereas systematic errors (such as out-of-calibration instrumentation or consistently low recovery rates during sample preparation) give rise to method bias. Working range will be of some interest in most cases. For trace work, limits of detection (LoD) and quantification (LoQ) may be relevant, but for planning calibration strategies it may be more useful to know the range over which the method response is linear.

3.3.1 Precision

The ISO definition of precision is 'the closeness of agreement between independent test/measurement results obtained under stipulated conditions'.[8] The important parts of this definition to note are, firstly, that the test results need to be independent, that is replicate measurements must be made following through the whole test method from sampling and sample preparation to the actual measurement procedure, and secondly, the precision is stated under defined conditions (usually either repeatability or reproducibility conditions, which are explained further in the following sections). Measurements for precision estimates should be performed on identical samples; the precision is a measure of how different the results are from each other in terms of both the spread from the highest to the lowest result point, and the distribution of results. It is important to note that the precision of the data obtained is not a reflection of its trueness (closeness to the true value), but is a measure of the random variability of results produced by the method. The relationship between precision, bias and accuracy is illustrated schematically in Figure 3.1.

An Introduction to Method Validation 45

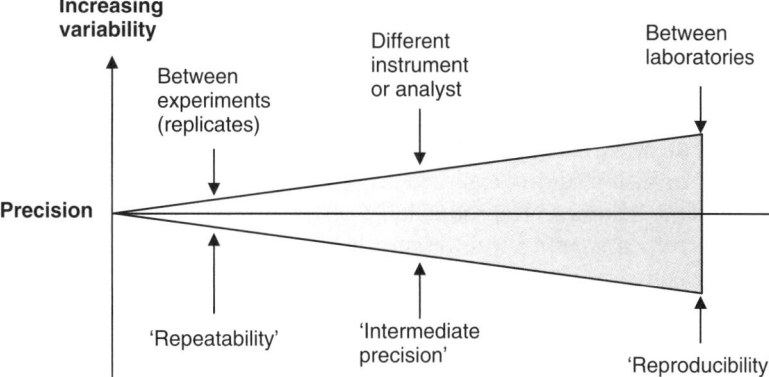

Figure 3.1 Schematic diagram illustrating the relationship between different precision estimates.

The usual statistical measure of precision is the standard deviation or relative standard deviation of the results obtained from the precision study. The relative standard deviation is defined as the standard deviation divided by the mean of the results, and can also be expressed as a percentage. The relative standard deviation is also referred to as the coefficient of variation (CV). The larger the CV or standard deviation, the larger the component of random error present in the system and the less precise the method is.

The conditions under which the repeated measurements are made will determine the type of precision estimate obtained. Three common types are repeatability, reproducibility and intermediate precision, described in the following sections.

3.3.1.1 Repeatability

Repeatability is the precision recorded from measurements made using a given method on a particular sample under similar conditions, for example the same analyst, same apparatus, same laboratory and a short interval of time occurring between analyses. These conditions are known as repeatability conditions. Repeatability represents the tightest extreme of independent precision measurements, and therefore indicates the variation that may typically be expected when measurements are made one after another by a single analyst. Knowing the repeatability of a method allows the analyst to judge whether differences in results obtained from different samples, analysed under repeatability conditions, are significant or merely due to random variation within the method.

For repeatability estimation it is important to minimise the time between repeat measurements, so for lengthy processes it is advisable to utilise other statistical approaches to pool data that are generated over a longer time period. These allow sufficient data to be produced to evaluate repeatability, without unwanted variation introduced from day-to-day effects on the method.

3.3.1.2 Reproducibility

Reproducibility is the precision recorded from measurements made using the same method on the same sample under differing conditions such as different analysts, using different machines, and working in different laboratories with long time intervals between analyses. Reproducibility represents the widest extreme of precision and is expected to reflect variation in the method from many possible sources. Reproducibility estimates are useful for deciding whether there is a significant difference between the results obtained for two samples measured in separate laboratories over an extended time period.

3.3.1.3 Intermediate Precision

Intermediate precision refers to the variation in results obtained in a single laboratory but over an extended period of time. It is generally used when a single laboratory uses multiple analysts and instruments for a particular method. The conditions for intermediate precision studies are specified by the laboratory (for example the study may involve a single analyst using different pieces of equipment, or may involve more than one analyst). It is therefore important to specify which conditions have been varied during the study. The aim of a study of intermediate precision is to estimate the likely variation in the results when the method is used routinely over an extended time period. The standard deviation obtained for intermediate precision for a particular method would be expected to fall between values obtained under repeatability and reproducibility conditions, as shown in Figure 3.1.

3.3.2 Bias

Bias is defined as 'the difference between the expectation of a test result or measurement result and a true value'.[8] Thus bias is a measure of the trueness of a result and is caused by systematic errors, rather than the random errors which influence the precision of results. In practice an 'accepted reference value' is normally substituted for the true value in the definition above. Bias can be evaluated experimentally by obtaining the average ($\bar{X}$) of a number of measurements of a sample with an accepted reference value (X_0), ideally a certified reference material with an associated uncertainty value (as shown in Figure 3.2). If a certified reference material is not available, as is currently the case for most DNA analyses, more readily available in-house samples, well characterised by other validated or reference methods, or spiked samples (prepared with a known amount of target material added) may be used. The reference material should match routinely analysed samples with regard to matrix composition, form and concentration of analyte as closely as possible, to ensure the bias measurements are meaningful across the full scope of the method, and to check that bias is sufficiently controlled across the full range of intended measurements.

Bias is calculated as the difference between the observed average value from the study and the accepted reference value of the test sample ($\bar{X} - X_0$), and is often expressed as a percentage difference from the expected reference value.

An Introduction to Method Validation

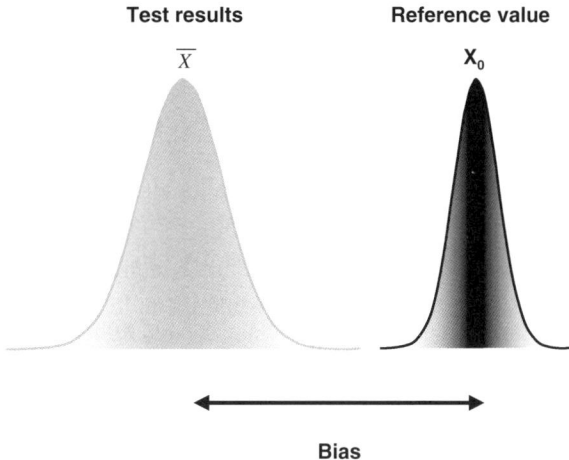

Figure 3.2 Schematic illustration of bias. $\bar{X}$ is the average of the test results, X_0 is the reference value and the bias is the difference between the two values.

Bias can also be expressed as a ratio ($\bar{X}/X_0$), or a percentage ratio. A *t*-test may then be used to determine if the measured bias is statistically significant. Statistical and practical biases are, however, not necessarily the same thing. In order to decide whether bias has practical significance for a specific method using particular conditions, the results of the *t*-test need to be looked at objectively in the analytical context.

A significant bias is often used as an indication that a method requires some further development or study. The bias may be due to unexpected effects operating through the method, or due to procedural errors or misinterpretation by the analyst. If significant bias is found, action is normally required to reduce it to insignificant levels. In some cases, however, the presence of bias can realistically be corrected for in the final result. The bias is not usually corrected across an entire analytical method, but may be corrected for specific effects such as temperature. One possible cause of bias is low recovery of the analyte, discussed in the following section.

3.3.3 Recovery

In many assays it is necessary to determine the overall recovery of the analyte achieved by the analysis. Applications such as the detection of pathogens in drinking water or the level of contaminants in foodstuffs require confidence that negative results truly reflect the absence of undesirable targets, rather than simply resulting from poor recoveries of the target. Spiked samples may be used to estimate percentage recovery through the extraction and quantification process, and are also useful in optimising sample preparation and analysis methods to maximise detection levels. In a biological context recovery usually refers specifically to the amount of added analyte that is recovered from a spiked sample. However, in chemistry it is often understood as the bias of the method expressed as a percentage. In practice, application of the analytical method is required to determine the level of recovery, which will be affected by any inherent bias in the measurement method, so determination of bias and recovery may be inextricably linked.

3.3.4 Accuracy

Accuracy is a property of a single result, and is defined as the 'closeness of agreement between a test result or measurement result and the true value'.[8] As in the case of the definition of bias, 'true value' is generally replaced by accepted reference value.

Accuracy is a combination of both the precision and the trueness of a measurement result. The true (reference) value is provided by a certified reference material or other sample as described for bias. A highly accurate method will produce results with a small bias which are also very precise; specifically the average of results is close to the true value and the spread of results is small. Conversely if a method is described as inaccurate, this would suggest a large bias (large difference between true value and average value of results), or imprecision (large spread of data), or both. This concept is illustrated in Figure 3.3.

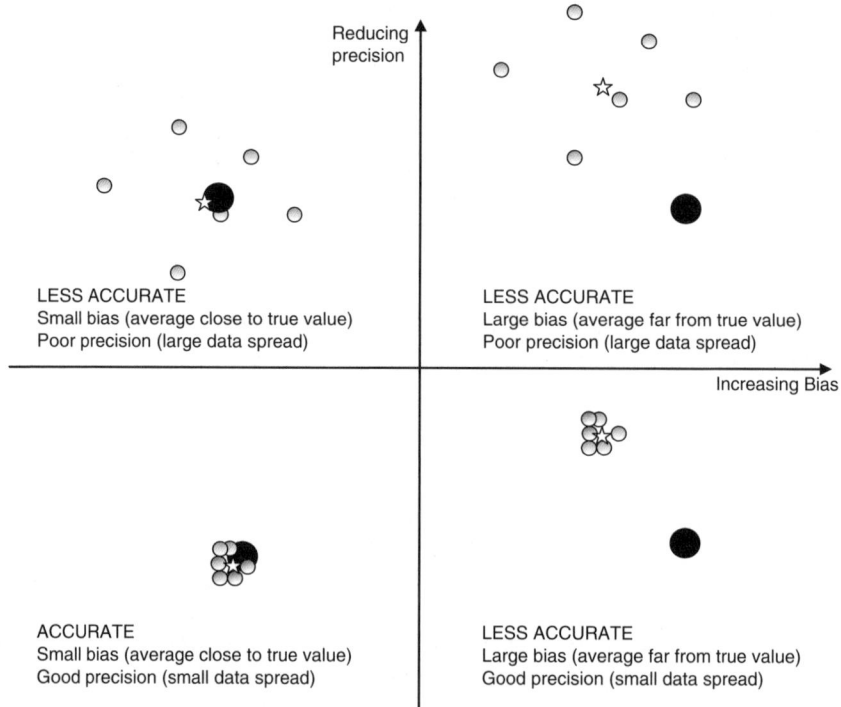

Figure 3.3 Schematic illustrations of accuracy and precision. Measured results are shown as shaded circles, average of results as a star and the true value (or accepted reference value) is shown as a large filled circle.

3.3.5 Ruggedness (Robustness) Testing

Experimental parameters will vary to some extent during routine operation of a method over time, and this may affect the method performance. Ruggedness testing, also known as robustness testing, helps to identify those parameters which have a significant effect on the performance of a method, and provides useful information on how closely such parameters need to be controlled to avoid the performance of the method being adversely affected. A rugged method is one whose performance is not affected by changes in the experimental parameters, within the defined control limits.

When evaluating ruggedness, experiments are designed which deliberately and systematically introduce known changes to parameters, and the effect on the result is assessed. One commonly used approach is that described by Youden and Steiner. This approach makes use of a 'Plackett and Burman' experimental design, which permits the study of up to seven experimental parameters through carrying out only eight experiments.[9] The changes introduced for each parameter should reflect likely variation which may occur during the normal operation of the method. Parameters identified as critical to method performance will vary depending on the method, but might include incubation times and temperatures, concentrations of buffer constituents, assay pH and volume of reagents added.

Parameters identified as having a significant effect on method performance require further study to identify suitable control limits. If the parameters cannot be controlled adequately then further development of the method will be required.

3.3.6 Selectivity

Selectivity is defined by IUPAC as 'the extent to which the method can be used to determine particular analytes in mixtures or matrices without interferences from other components of similar behaviour'.[10] Selectivity and specificity are often used synonymously; specificity can be termed the ultimate in selectivity as if a process is specific it is by definition wholly selective.

The selectivity of an analytical method may be affected by many factors, including the presence of impurities, degraded components and possible inhibitors or enhancers of the reaction, and physical parameters such as temperature, ionic strength or pH. The design of selectivity experiments therefore requires both background technical knowledge and information about the typical samples which will be analysed using the method. An ideal selectivity experiment should test the possible effects of all possible interferents on the typical observations, although this is rarely possible in practice.

Poor selectivity of a reaction indicates that other substances can interfere with the analysis, and selectivity may change as assay conditions are altered. For example, a PCR at the optimised annealing temperature will produce a single, specific amplification product but, if the annealing temperature is changed and the reaction made less stringent, multiple non-specific products may be generated.

In clinical diagnostics the selectivity of an assay is often reported, and is calculated as the percentage of true negative results obtained in testing a number of known negative samples.

3.3.7 Detection Limit (Sensitivity)

It should be noted that the biological definition of sensitivity given here varies from the ISO chemical definition. Detection limit terminology is inconsistent and confusing, so it is important to try and follow specific sector guidelines where available. In most biological analysis, sensitivity is used to describe the lowest level of an analyte that can be measured. However, in fields outside of biological analysis the term 'sensitivity' has different definitions. For example, in chemical analysis, sensitivity is usually defined as 'the change in the response of a measuring instrument divided by the corresponding change in the stimulus'.[11] In clinical applications sensitivity again has a slightly different meaning, and is often expressed as the percentage of tests that give the correct positive result in testing a number of known positives.

Often two limits are defined for a quantitative assay, firstly the limit of detection (LoD), which is the lowest amount of a target which can be reliably detected and distinguished from zero results and background signals with confidence. The second is the limit of quantification (LoQ), which is the lowest concentration of analyte that can be quantitatively measured with an acceptable level of uncertainty.

Practically there are several ways of determining the sensitivity of a method. In qualitative analysis the analyte is typically diluted serially until it can no longer be detected reliably using the method (usually once the percentage detection falls below a specific level, often 95 or 100%). A variety of approaches is used in quantitative determinations, but most approaches use the results of repeated analysis of a negative sample or zero calibrator (a sample known not to contain the analyte of interest). The zero calibrator is analysed between 10 and 20 times, and the mean and standard deviation of the data obtained. Usually the limit of detection, or analytical sensitivity, is set as the mean signal +2 standard deviations while the limit of quantification is set as the mean +10 standard deviations. In quantitative PCR analysis, for example, where negative results do not yield a meaningful value, the definition of LoD and LoQ is more difficult. One developed approach is to define the LoD as the input analyte level giving a (for example) 95% probability of a positive PCR result, calculated using probit regression analysis of dilution series data.[12]

Sensitivity can be expressed in many ways depending on the assay, for example:

- Number of cells per mass of matrix that is detectable;
- Percentage of adulterant in a matrix that is detectable;
- Mass of DNA required for reliable qualitative analysis (such as STR profiling);

- Amount or copy number of a gene, genome or DNA target that is detectable, per volume or mass.

Sensitivity should be assessed across the range of analyte levels and sample masses that may be routinely used, as the ratio of detectable target within a given sample cannot always be linearly extrapolated. For example if 100 microbial cells can be detected in 1 gram of soil by PCR, 10 cells may not be detectable if only 0.1 g of soil is tested.

Sensitivity is often determined as part of a method validation, and gives an indication of the lower operating limits of the test method. Most experiments to determine detection limits require a replicated sequence of experiments on low-level samples, blanks and low-level spiked materials or standards. Where the results of an analysis are often close to the lower operating limit of a method it is advisable to perform regular assessment of the LoD. This ensures that negative results are being interpreted and reported correctly with reference to the detection or quantification limit of the method.

3.3.8 Working Range and Linearity

The working range is the interval between the upper and the lower concentration of an analyte in the sample for which it has been determined that the method is suitable; generally this is the range where the results have acceptable uncertainty (Figure 3.4). The upper boundary of the working range is defined by the concentration at which there is insufficient change of response per unit of concentration, and may reflect the upper limit of the instrument or the exhaustion of one or more reagents in the assay. The lower limit of the range is usually determined by the limit of quantification, beyond which results cannot be determined with an acceptable uncertainty.

The term linearity is frequently linked with the working range of the method, and refers to the ability of a method to give a response directly proportional to the concentration of the analyte. The linear range of a method is, therefore, the

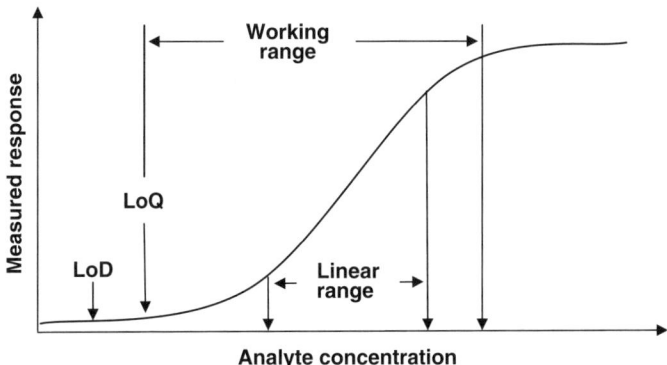

Figure 3.4 Illustration of some key performance characteristics: Working range, linear range, LoD and LoQ.

range over which the response is directly proportional to concentration, and can be smaller than the working range.

In order to establish the working range of a method it is necessary to study the response of standards whose concentrations span the range of interest plus 10–20% outside the range. The standards should be evenly spaced across the range to avoid introducing any bias into the experiment. Obtaining standards for the study of linearity and working range may be difficult as certified reference materials (CRMs) suitable for molecular analyses are not widely available. An exception is soya flour CRMs with known genetically modified (GM) soya content which have been certified on a mass-per-mass basis, and are used to underpin quantitative GM analysis of certain foodstuffs.

In the absence of commercial standards, in-house samples or spiked samples well characterised by other methods may be used. Typically a dilution series of the well-characterised sample is used to assess the linearity of a method. Care should be taken to minimise analytical error in preparing the dilutions. Linearity should be established for each matrix specified in the scope of the method, and the working range of any equipment used must also be assessed.

In order to evaluate linearity, the results should first be plotted and a visual inspection carried out to identify any obvious deviations. Subsequent statistical analysis, including determination of the correlation coefficient (r), residual standard deviation, standard deviation of the slope and intercept and residuals plots, is then performed to demonstrate objectively the fitness for purpose of the method with respect to linearity.

3.3.9 Measurement Uncertainty

Measurement uncertainty can be defined as an estimated range of values within which the true value of the measurement lies. The range of values gives an indication of the reliability of a measurement result. The experimental result may then be reported as $x \pm y$, where x is the reported measured value and $\pm y$ is the degree of uncertainty associated with the measurement result.

Experiments performed in method validation often provide information which can be used in evaluating measurement uncertainty. In the assessment, all possible sources of error in the measurement process are considered and evaluated, to create an uncertainty budget. A statement of measurement uncertainty should not be assumed to imply doubt about the validity of the result; on the contrary, the presence of an uncertainty value should increase confidence in the validity of the results. A statement of the uncertainty enables both the analyst and other end users of the experimental data to assess the reliability of the result, and to judge the level of confidence that can be placed in any decision based on the result. Measurement uncertainty, when evaluated correctly, will give a more realistic estimate of the true spread of results, compared to a repeatability estimate. An uncertainty estimate should take account of all the factors in the measurement process that have a significant effect on the measurement results, and will therefore include the effect of both random and systematic errors. In contrast, estimates of precision (such as repeatability) only include the effects of random error.[13,14]

3.4 Validation in Practice

To facilitate practical application of the theoretical principles described so far, a broad outline procedure is given in the following sections, together with an example of the data required for validating an assay using quantitative PCR (qPCR) analysis for environmental testing.

The assay in this example is a qPCR method for the detection of a bacterial pathogen. The assay is technically challenging because of possible inhibitors in the environmentally derived samples, and the pathogen may be present in very low amounts, yet still pose a threat to human health. A published method is being introduced to a new laboratory, and evidence of the fitness for purpose of the method in the new environment is required.

This example is a simple situation requiring a single analytical technique to be used to determine the analytical result, namely direct qPCR measurement of the sample. In practice an analysis is typically more complex, requiring a number of sampling, preparation and purification or extraction stages in addition to the analysis, all of which may require validation.

3.4.1 Outline of the Procedure

There are various logical stages in undertaking a method validation process, although the level of detail and effort required should match the importance and/or the amount of subsequent use of the methodology.

3.4.1.1 Define the Analytical Requirement

The first step in developing a method that is fit for purpose is to understand fully the requirements of the analysis. Information such as the analyte to be tested, the analytical matrix and whether qualitative or quantitative information is required will all help define the aim of the assay. It is also important to know why the measurements are being made and what the results will be used for, in order to judge correctly the criticality of the data. For example, the performance requirements for a method which will be used to assess whether a legal limit has been exceeded are likely to be much more demanding than for a method used to 'screen out' samples that will be sent for further testing.

In this example, the customer has requested that as few as 100 organisms must be detected per test sample, and a quantitative result is required with a repeatability CV of less than 5% and a bias of less than 15%. The results will be used to determine if the tested water sources have above the maximum level of the bacterial pathogen. The information provided from the test will be used to determine the fitness of drinking water for human consumption. Thus a high level of accuracy and confidence in the results is required.

Samples will be provided as 10 ml purified extracts from drinking water concentrates, but will not have been tested for the presence of inhibitory substances.

3.4.1.2 Write a Draft Protocol

In this example, the method used to set up and perform the quantitative PCR measurement of prepared water samples is first detailed in the draft protocol. The process is based on the published assay, but details of the in-house instruments and reagents that will be used for the assay are included.

3.4.1.3 Investigate the Robustness of the Technique, and Identify the Critical Parameters

Critical parameters in the method may already have been stated in the published method. If the information is not available, experiments altering a range of qPCR parameters including the amount of reaction components that are added to the reactions, the annealing temperature and the thermal cycling conditions can be carried out to determine the experimental variables that are most likely to affect the overall performance of the assay. The protocol can then be amended to define the level of control required for any critical stages identified.

3.4.1.4 Identify Relevant Performance Parameters, and Determine the Order of Investigation

For the qPCR analysis in this example, relevant performance characteristics would be the specificity of the test method, the working range (also often called the linear dynamic range in qPCR terminology) and sensitivity of the assay. Estimates of bias in the measurement results, the repeatability and the measurement uncertainty are also required. A logical order for assessing the required parameters would be to begin with the specificity of the method, ensuring that only the pathogen of interest yields positive results. Subsequently sensitivity, linear range, repeatability and bias can be determined. This allows the expected performance characteristics of the method to be established, with defined acceptable ranges for the repeatability, bias, linear range and accuracy. Finally an uncertainty budget, based on the collected performance data may be developed.

3.4.1.5 Assess the Performance Characteristics Using Suitable 'Known' Materials (RMs, Standards, Spikes)

Materials for assessing the specificity of the assay, such as other microbial species that may also be present in the environmental samples or related organisms that might cross-react, may be obtained from a range of sources including culture collections. Homogenous aliquots of the target may be used to test the precision of the assay. However, choosing suitable materials to

An Introduction to Method Validation

determine the bias of the method is more difficult, as appropriate quantitative reference materials are not available for the majority of DNA-based methods. Use of several different methods or laboratories to characterise a sample, by reaching a consensus value which may then be used to assess method performance, is a possible approach. In the case of bacterial testing, samples may be split and a portion tested by DNA extraction and qPCR quantification while the concentration of bacteria in a second portion is determined through serial dilution and plate counting. This allows the agreement of the results produced by the test method with the actual number of organisms present in the sample to be evaluated.

In our example, five unknown samples were tested by eight independent expert laboratories, and consensus values were obtained for the number of organisms in each sample. The characterised samples were then used to assess the performance of the in-house qPCR method.

3.4.1.6 Assess Whether the Data Show the Method is Fit for Purpose

The results obtained through the planned testing are then assessed statistically, with key results in this example being the sensitivity achieved and the precision and bias of the method, identified from the original customer requirements.

Firstly DNA samples from a range of water-borne pathogens and related bacterial species were used as the analyte, at a concentration of 10^4 organisms per reaction, to determine the selectivity of the method. Positive signals were only detected from the target organism, demonstrating assay selectivity. Secondly a dilution series of target DNA from the pathogen, ranging from 1 to 10^9 organisms (also referred to as genome equivalents), was prepared and run in triplicate using the qPCR assay. The measured Ct values (see Chapter 7) were plotted to assess the range over which the relationship between the analyte concentration and measurement were linear (Figure 3.5).

The error bars show the standard deviation of the triplicate measurements at each dilution.

Below ten organisms per reaction the Ct repeatability, as assessed by the CV, increases above the required limit of 5% for the assay. Beyond this dilution the samples are not measured with sufficient precision, as the variability increases because of greater sampling variability at the low target concentration. This is taken as the limit of quantification (LoQ) of the assay. Similarly, below five organisms the sample is not detected with sufficient reliability (as less than 95% of replicates at this level yield a positive signal), and this is the limit of detection of the method. The required LoQ of the method is 100 organisms per sample. However, only one fifth of the original 10 ml sample can be analysed in one reaction post-extraction, as the sample is concentrated into a 50 μl DNA extract, 10 μl of which is added to each PCR. This means that the required LoQ per qPCR reaction is actually 20 organisms, which is achieved in the evaluation of the working range and sensitivity undertaken here.

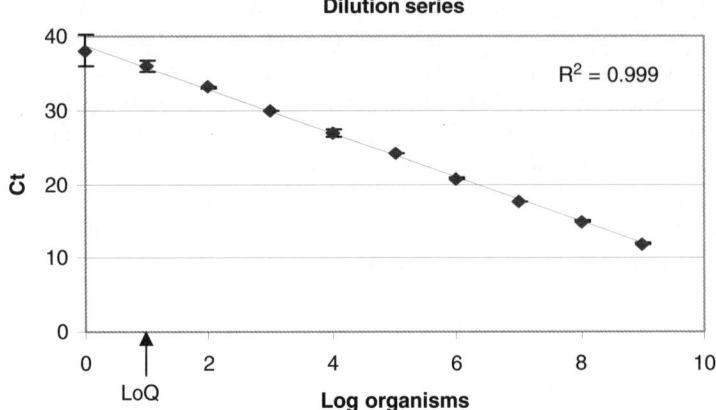

Figure 3.5 Establishment of linear range, by performing the bacterial pathogen qPCR assay on a dilution series of target DNA.

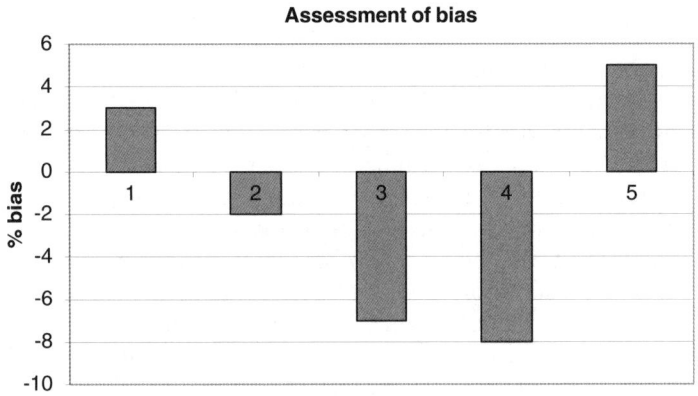

Figure 3.6 Plot showing the percentage bias of the result of test samples analysed using the qPCR assay.

Thirdly the bias (Figure 3.6) and precision (Figure 3.7) of the assay were evaluated by performing six replicates of the assay on each of the five samples characterised by the inter-laboratory analysis. The bias was determined using $\bar{X} - X_0$, where $\bar{X}$ is the average of the measured values for each sample and X_0 is the consensus value of the sample.

The assay was performed under repeatability conditions to derive six sets of results for each of the five characterised samples. The results were analysed to determine the repeatability of the method. The measured Ct values were used for this assessment, and the %CV was calculated as the standard deviation of each set of six measurements divided by the mean and multiplied by 100 (Figure 3.7).

The initial stated requirements were for a reliable assay specific for the target pathogen, with an LoQ of at least 20 organisms per reaction, a repeatability CV

An Introduction to Method Validation 57

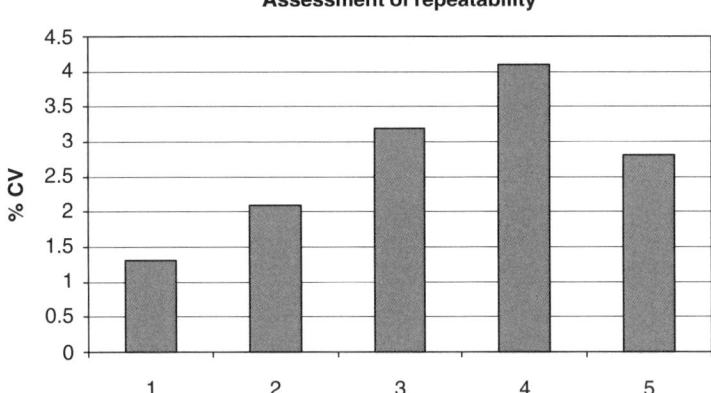

Figure 3.7 Repeatability of results using the test method on five well-characterised samples, evaluated using %CV.

of less than 5% and a bias of less than 15%. The performance characteristics of the test method meet these specifications.

3.4.1.7 Define the Limitations of the Methodology

If the assay is affected by the presence of certain other environmental bacteria, for example, then this should be noted in the final scope of the method.

3.4.1.8 Document the Final Protocol and Method Validation Results

The last stage of the validation process is to record the details of the validation procedure and the results of the performance characteristics investigation. This information should be made available with the final method protocol and details of the critical parameters of the method, together with the acceptable ranges of performance defined during the validation.

Acknowledgements

The authors would like to thank Malcolm Burns for helpful discussions. By kind permission some of the information and diagrams have been reproduced from the LGC training course on Method Validation.[15]

References

1. ISO/IEC 17025:2005, *General Requirements for the Competence of Testing and Calibration Laboratories*, International Organization for Standardization, Geneva, 2005.

2. EURACHEM, *The Fitness for Purpose of Analytical Methods: A Laboratory Guide to Method Validation and Related Topics*, 1998, ISBN 0-948926-12-0.
3. LGC, *In-House Method Validation. A Guide for Chemical Laboratories*, 2003, ISBN 0-948926-18-X.
4. LGC, *mVAL Software for Analytical Method Validation*, 2003.
5. M. Thompson, S. L. Ellison and R. Wood, *Pure Appl. Chem.*, 2002, **74**, 835.
6. GLPMA. Department of Health (2000). The United Kingdom Good Laboratory Practice Monitoring Authority guide to UK GLP regulations 1999, available at:http://www.mhra.gov.uk/home/idcplg?IdcService=GET_FILE&dID=1306&noSaveAs=0&Rendition=WEB (accessed July 2007).
7. ISO 9001:2000, *Quality Management Systems – Requirements*, International Organization for Standardization, Geneva, 2000.
8. ISO 3534-2:2006, *Statistics – Vocabulary and Symbols – Part 2: Applied Statistics*. International Organization for Standardization, Geneva, 2006.
9. W. H. Youden and E. H. Steiner,*Statistical Manual of the AOAC*, AOAC International, 1975, ISBN 0-935584-15-3.
10. J. Vessman, R. I. Stefan, J. F. van Staden, K. Danzer, W. Lindner, D. T. Burns, A. Fajgelj and H. Müller, *Pure Appl. Chem.*, 2001, **73**, 1381.
11. ISO, *International Vocabulary of Basic and General Terms in Metrology*, International Organization for Standardization, Geneva, 1993, ISBN 92-67-101751.
12. C. Drosten, E. Seifried and W. K. Roth, *J. Clin. Microbiol.*, 2001, **39**, 4302.
13. Eurachem/CITAC Guide, *Quantifying Uncertainty in Analytical Measurement* 2nd edn, EURACHEM, 2000, ISBN 0-948926-15-5.
14. ISO, *Guide to the Expression of Uncertainty in Measurement*, International Organization for Standardization, Geneva, 1995, ISBN 92-67-101889.
15. LGC, *Analytical Quality Training Programme, Method Validation Training Course*, 2007.

CHAPTER 4
DNA Extraction

GINNY C. SAUNDERS AND JENNIFER M. ROSSI

LGC, Queens Road, Teddington, TW11 0LY

4.1 Introduction

The isolation of genomic DNA is a fundamental requirement for many analytical molecular biology procedures. Although proper collection and stabilisation of the sample is crucial, purification of the DNA is often the key step for success in downstream analysis. This chapter is concerned with situations where DNA extraction forms the first step in the analytical process. Without the successful execution of this step, further analysis using detection and identification techniques (for example, PCR detection of specific genes or mutations, DNA sequencing and microarray analysis) would be severely compromised.

In recent years the use of commercial DNA extraction kits has become more commonplace, both to avoid the use of hazardous reagents and to choose ready optimised procedures for the specific sample type. Benefits include the potential for more consistent performance, no requirement for the use and disposal of hazardous solvents and other reagents and reduced optimisation times as an appropriate kit can often be purchased specific for the sample type. However, the use of commercial kits is often more expensive than in-house methods, although recently a method for recycling of commercial columns to reduce the associated costs has been published.[1]

In addition, automated DNA extraction is being more widely adopted as laboratory throughput and assays requiring efficient DNA extraction have increased. Regardless of the DNA isolation technique employed, a wide range of instrumentation is now available for low- to high-throughput processing of samples. Automation of the extraction process frees up technician time for additional tasks and enables more standardised DNA preparations in the laboratory, by reducing human error and streamlining methods.

In cases where a sample is in limited supply (such as archival material) or even unique (such as scene-of-crime forensic samples), successful extraction of the valuable DNA analyte is of paramount importance. Equally, even if the sample is in plentiful supply, a well-characterised and robust DNA extraction technique is required that will reproducibly isolate DNA within well-defined detection limits, and that is appropriate to the sample matrix or organism. Often, the sample matrix may include components inhibitive to the downstream assay, making robust DNA extraction even more important.

The variety and complexity of samples submitted for analysis using molecular biological techniques is vast. It is therefore beyond the scope of this chapter to offer a complete range of validated DNA extraction protocols. Equally, a small selection of protocols would have no more than a very limited relevance. Well-developed protocols are perhaps most appropriately obtained from molecular biologists specialising in the sample or field of interest, often published in dedicated molecular biology manuals or journals specific to the particular field of study. Sources of a selection of DNA extraction techniques are given in Section 4.6. The information presented in this chapter aims to highlight generic validation issues that can arise at various stages in a broad range of DNA extraction procedures. In-house evaluation and optimisation of the relevant steps are necessary for each sample or matrix type to ensure reliability and validity of the methods.

Ideally, in an analytical environment, an effective DNA extraction procedure should be as simple, safe and cost and time efficient as possible. It should also reproducibly provide DNA of a sufficient quality and yield to allow subsequent analysis.

The suitability of isolated DNA as an analyte for a given technique is generally determined by three important factors: (i) amount or concentration, (ii) purity and (iii) integrity of the DNA. Each of these factors can be influenced by the extraction technique employed and, in turn, impact upon the validity of techniques applied in subsequent analysis.

4.1.1 Concentration or Amount

The amount of DNA obtained must be sufficient for all analyses, including relevant controls and duplications, to be carried out. The DNA must also be available at a workable concentration. Optimisation of the scale of the extraction technique, with respect to the required amount of analyte, will offer optimum reagent consumption and therefore better value for money.

4.1.2 Purity

The isolated DNA should be as free as possible from all contaminants, both endogenous and exogenous, that may inhibit subsequent analyses. Where possible and appropriate, all contaminating nucleic acids should also be removed.

4.1.3 Integrity

High molecular weight (HMW) DNA (ranging from 50–200 kb) can be a requirement for some types of analyses, whereas in other cases a degree of

DNA degradation can be tolerated. If the DNA is sheared or degraded into fragments, the extent of such damage may be determined by agarose gel electrophoresis, in order to confidently employ the most suitable method of analysis.

4.2 Steps of the DNA Extraction Process

A generic DNA extraction process has to achieve a number of specific aims, set out below. In practice, an extraction technique may not necessarily consist of the sequential execution of each task. For example, extraction buffers commonly contain lysis chemicals as well as chemicals to chelate inhibitors. Depending on the sample type and subsequent analysis, greater or lesser emphasis may be placed on each of the following steps as required.

4.2.1 Sample Preparation

A sample may benefit from a number of preparative steps prior to cell or membrane lysis. This could include homogenisation, centrifugal separation or a step to minimise the effects of surface contaminants.

4.2.2 Cell or Membrane Lysis

This step disrupts the cell wall/membrane and frees the DNA from cellular and organelle membranes. This can be accomplished by chemical (usually detergents), mechanical, enzymatic, microwave, sonication, heat or freeze/thaw treatment.

4.2.3 Protection and Stabilisation of Released DNA

An extraction buffer is usually present during the lysis process. This contains a combination of chemical components which protect the released DNA in its new environment from degradation by cellular nucleases liberated during lysis.

4.2.4 Separation of Nucleic Acids from Cell Debris or Sample Matrix

The separation of released DNA from cellular and matrix debris and other biological macro-molecules has traditionally been achieved by phenol:chloroform and chloroform extractions, although commercial methodologies are now routinely used which do not require the use of such hazardous reagents.

4.2.5 Purification of DNA

RNA may be removed from crude DNA extracts by the addition of appropriate nucleases. Other impurities that may act as inhibitors can, in some cases, be removed by appropriate ion exchange columns, bead-based chemistries or the addition of chelating agents.

4.2.6 Concentration of DNA

Both alcohol precipitation and commercial columns or bead-based chemistries can concentrate the DNA to a suitable working molarity. Some inhibitors may also be removed at this stage.

Commercial kits are available which contain all necessary reagents for sample preparation, cell and membrane lysis, protection and stabilisation of DNA and separation of nucleic acids from cell debris or sample matrix. Kits are generally based on silica columns or magnetic beads over a wide range of chemistries.

4.3 Choosing an Appropriate DNA Extraction Procedure

Before a DNA extraction process is undertaken on a sample type for the first time, it is essential for the analyst to be informed of, and understand the consequences of, the following characteristics of the sample and the requirements of analysis.

4.3.1 The History of the Sample

- Is the sample unique? If so, DNA yields may need to be maximised to allow repeat or multiple analyses;
- Has the sample been maintained in a stable environment? This may help determine whether a method optimised for HMW or degraded DNA should be employed;
- Has the sample been exposed to a possible source of contamination? If so, certain sample preparation steps may be required that will remove contaminants.

4.3.2 The Composition of the Sample

- Is the sample heterogeneous with respect to the analyte and matrix components? If so, preparation steps may be required to homogenise the sample or sampling issues should be addressed;
- Is the analyte contained in a matrix whose components are well characterised with respect to potential inhibitors or extraction strategies? If so, an existing extraction process could be adapted and optimised;
- Do the target organisms have cellular structures that lyse sufficiently under the conditions to be employed? This should be determined empirically;
- If multiple target organisms or tissues are being analysed simultaneously, will the lysis method used lyse all cells with equal efficiency? This is

important if multiple target quantification or detection is required from a single extraction, and should be determined empirically (for example, Gram-positive and Gram-negative bacteria may lyse with different efficiency under the same conditions);
- Should the number of nuclei per mass of tissue or organism (and thus the theoretical yield) be taken into account if quantification is required? This should be determined empirically.

4.3.3 Time and Resources Available

- Should the sample be analysed by employing an extensive and possibly lengthy extraction process in order to maximise yield, purity and integrity or, if time is of the essence, can these factors be selectively compromised and the sample adequately analysed by employing a shorter, somewhat cruder technique?
- Is a high throughput of samples required? If so, a technique with minimum hands-on time might be more appropriate if it does not compromise analytical quality. Additionally, automation of the extraction process could be considered.

4.3.4 Standardised Techniques

- Are results of multiple technicians or laboratories going to be compared? If so, a standard method should be chosen for use;
- Will results from a wide variety of sample types be analysed simultaneously and compared (for example, forensic scene-of-crime samples)? If so, standardised commercial kits with specialised protocols or instrumentation platforms should be considered.

4.3.5 Subsequent Analytical Procedures

- Could endogenous chemicals (such as secondary metabolites) or exogenous compounds (such as food or fibre dyes) be present that inhibit or interfere with certain types of analyses? If so additional purification steps may be required;
- Is quantitative analysis required? If so, appropriate steps to ensure sample homogeneity should be undertaken;
- Is HMW genomic DNA required or will degraded DNA be tolerated as an analyte? (See Table 4.1.)

4.3.6 Potential Impact of Methodology

An example of some of the points discussed in this Section is given in Figure 4.1, where DNA was extracted from meat samples. At the stage prior to DNA

Table 4.1 DNA requirements of some common analytical techniques that employ genomic DNA.

Analytical method	DNA requirements		
	Purity	Integrity	Approx. amount
Gene specific PCR	High-crude tolerated	HMW or some degradation tolerated	1–5 ng µl^{-1}
RFLP fingerprinting	High	HMW required	20–100 ng
AFLP profiling	High	HMW but some degradation tolerated	10 ng/reaction
Dot or slot blot hybridisation	High	HMW but some degradation tolerated	100 ng–10 µg/dot
PFGE	High	Very HMW required > 50 kb	1–10 µg/sample
Cycle sequencing	High	Degradation tolerated	10 ng µl^{-1}
STR analysis	High	Degradation tolerated	0.1 ng µl^{-1}
Real-time, quantitative or multiplex PCR	High	Some degradation tolerated	0.1–5 ng µl^{-1}

precipitation, the extract was divided into four and precipitated as described (in the figure legend) before an aliquot was electrophoresed on an agarose gel and visualised under UV light. Figure 4.1 demonstrates how the alteration of a single procedure in the extraction process can alter the result obtained. In this instance, one of the four alcohol precipitations carried out (Lane 3) significantly reduced the yield of the DNA obtained. By comparing Lanes A1, B1, C1 and D1 it can be seen that different DNA yields are obtained from the different tissue types, with reduced yields being associated with higher fat content samples. This could be due to differing cell lysis efficiencies or different cell size and therefore the number of nuclei present. It can also be seen that different quality of DNA is obtained; the DNA obtained from the kidney tissue (B) shows signs of degradation, typically associated with programmed cell death. This could be due to poor storage conditions or tissue specific enzyme activity.

Choice of appropriate methodology can thus have a significant impact on the yield and quality of extracted nucleic acids available for downstream analyses. In situations where analysis is confined to a single type of sample such as fresh blood, this decision may only have to be taken once, with the method validated

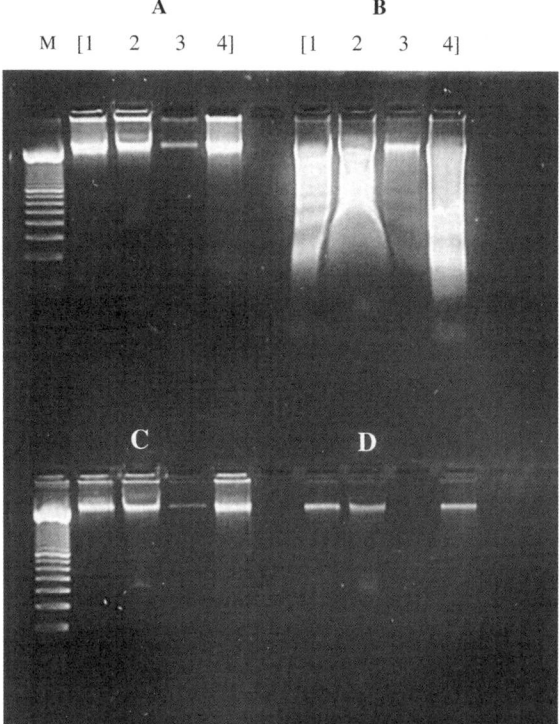

Figure 4.1 Precipitation of DNA obtained from different mammalian tissue using standard ethanol and isopropanol procedures. Equal amounts of four types of bovine meat tissue, A, steak; B, kidney; C, stewing steak; D, fat, were extracted using an optimised and validated method for meat samples. Lane 1, DNA precipitated with ethanol and ammonium acetate. Lane 2, DNA precipitated with ethanol. Lane 3, DNA precipitated with isopropanol and ammonium acetate. Lane 4, DNA precipitated with isopropanol. M: 100 bp molecular weight marker.

and reviewed as the technology develops. However, some analytical laboratories are faced with the challenge of analysing DNA extracted from many different types of matrices. For example, the detection of adulteration in foodstuffs may involve identification of different meat species in various cooked products, the identification of different wheat species in dried pasta or the detection of animal matter in a dehydrated soup or sauce mix. In these cases the characteristics of the sample and analysis required, as discussed above, would have to be taken into account for each individual sample type before the most suitable extraction technique could be identified.

Under these circumstances it may be useful to take a unit-based approach to the DNA extraction process, breaking down the procedure into smaller, 'stand-alone' units that could be validated in isolation. By applying a 'mix and match' selection of the most appropriate and efficient units of the extraction process, a tailor-made extraction process can be established. This approach can, when applied with due consideration and knowledge, reduce the need to re-validate an entire DNA extraction technique for each unique sample.

4.4 Validation Issues Arising at the Various Stages of DNA Extraction

The unit-based approach will also be adopted in this chapter for the evaluation of validation issues concerning DNA extraction techniques. Factors most likely to affect the yield, quality (purity and integrity), reproducibility and overall robustness of the various stages of an extraction process are discussed.

4.4.1 Sample Storage

Storage of the sample before submission for analysis can often be outside of the control of the analyst. Once in the laboratory, samples should be immediately placed in a stable and suitable environment. Such precautions will prevent contamination and further degradation of the sample. A number of factors have been identified which may affect the validity of DNA extraction due to sample storage, and are detailed in the following sections.

4.4.1.1 Incorrect Sample Storage Temperature

Further growth of contaminating micro-organisms can be halted by freezing a sample. Naturally occurring autolysis and DNA degradation by endogenous and exogenous enzymes can also be abated by reduced temperatures. Samples requiring long-term storage should either be placed at –80 °C, preferably after snap cooling (fast freezing in liquid nitrogen), or lyophilised,[2] after which a sample can be stored at room temperature in a dry environment for extended periods. Any drying or freezing procedures should be undertaken as soon as possible after the arrival of the sample. Excessive freeze/thaw treatment of a sample should be avoided, as this may induce cell or DNA breakdown. Suitable precautions at this stage can help maintain the integrity of the DNA.

4.4.1.2 Incorrect Sample Storage Environment

Humidity can encourage microbial growth. Dust and aerosols originating from other samples can result in cross-contamination and the production of false positive results. Storage in a sterile, dry, sealed and sometimes dark environment may be necessary for samples placed at ambient or reduced temperatures.

4.4.2 Sample Preparation

A sample may require some manipulation in order to maximise the ratio of surface area to lysis forces. In some circumstances it may be prudent to surface-clean or sterilise a sample, for example if the target organism is contained within a matrix or higher organism which could have been externally contaminated. Selective separation of the sample could also take place at this stage. Factors affecting the validity of DNA extraction due to sample preparation have been identified as follows:

4.4.2.1 Homogeneity of Sample

It is essential that material selected for analysis is truly representative of the entire sample with respect to both matrix and analyte.[3] In order to achieve this, it may be necessary to homogenise heterogeneous samples in a mixer mill, blender or mortar and pestle before analysis. Alternatively, a minimum sample mass or size should be employed, thereby reducing the amount of misrepresentative sampling that may occur if a much smaller heterogeneous selection is made. Adherence to such empirically determined criteria can significantly improve reproducibility of yield and accuracy in quantitative analysis and avoid the possibility of false negative results.

4.4.2.2 Surface Area to Lysis Forces Ratio

Bulky samples such as animal tissue may require breaking down into smaller segments or homogenisation in order to increase the efficiency of cell lysis. Failure to undertake this measure or to carry it out in a consistent manner may result in low or non-reproducible yields of DNA.

4.4.2.3 Cell or Nucleic Acid Adherence to Matrix Material

There is sometimes a requirement to remove cells from complex matrices or nuclei from cells before cell or membrane lysis takes place. In this way, inhibitors present in the matrix or cell are removed at the beginning of the analysis. Caution should be exercised, as a percentage of cells and DNA may adhere to, or be entrapped within, certain matrices. This can certainly be the case with soil aggregates and some soil bacteria or free DNA. Consideration and avoidance of analyte loss can maximise yield, improve accuracy of quantitative analyses and maintain homogeneity of sampling.

4.4.2.4 Contamination

Extreme care should be exercised to ensure that cross-contamination of samples does not occur during the early stages of analysis. Such occurrences could happen when a blender is used to homogenise multiple samples without adequate cleaning between specimens. Alternatively, the production of fine dust particles when grinding materials with liquid nitrogen should be suitably contained to avoid corruption of other samples. Protective clothing and gloves should be worn by all analysts to prevent bodily fluids or skin scales being inadvertently introduced into a sample and confusing the analysis.

4.4.3 Cell and Membrane Lysis

Cell lysis and the maintenance of DNA integrity are often conflicting aims of this part of the technique and sometimes a compromise has to be struck between maximising lysis activity and minimising shearing or degradation of the DNA. Several approaches can be taken for releasing nuclear DNA. The most common method is when both cell and membrane lysis is accomplished simultaneously under the same conditions.

When working with fresh plant and fungal material, it is sometimes preferable to achieve cell wall degradation first, followed by DNA isolation from protoplasts. In situations where inhibitory interference from cytoplasmic contaminants is to be avoided, intact nuclei can be isolated from cellular debris prior to lysis of the organelle or nuclear membrane. Although the last two approaches can be time consuming and are not generally suited to an analytical environment, they allow the genomic DNA to be released into an environment that minimises the degradation of the analyte and is relatively free of inhibitors.

A common cause of DNA degradation is the activity of endogenous DNases. The conditions for optimum activity of these nucleases vary from species to species, however a number require magnesium as an essential co-factor (others may be dependent on other metallic co-factors even when present in the same organism,[4] and will only be active within a limited pH range. Addition of EDTA to the extraction buffer chelates magnesium ions, thereby reducing the chances of nuclease-induced DNA degradation. Optimisation of the extraction buffer pH and the addition of detergents may also minimise endogenous DNase activity. Oxidative damage of DNA can also occur, particularly if samples are stored dried.[5]

Factors affecting cell or membrane lysis have been identified as follows:

4.4.3.1 *Inaccessibility of Cells to Lysis Forces*

If cells are not available to the enzyme or detergent, cell lysis will be inefficient, reducing yield and reproducibility (see Section 4.4.2).

4.4.3.2 *Type and Amount of Detergent or Denaturant Used*

There are many different types of detergents and denaturants available for cell or membrane lysis (Table 4.2). Generally a detergent binds to the membrane and membrane lysis and solubilisation occurs to give detergent–lipid–protein complexes. Several factors can affect the performance of a given detergent including temperature, pH, ionic strength, detergent concentration, presence of multivalent ions and the presence of organic additives.[6] Therefore, if any of these factors are altered in a standard protocol, the effect on detergent activity should be empirically determined.

4.4.3.3 *Concentration and Activity of Lytic Enzyme*

Enzymes also perform optimally within a specific range of pH and temperatures, and thus may be affected by the presence of metal ions or chelating

Table 4.2 Common detergents and denaturants used in DNA extraction.

Detergent/denaturant	Description	Final concentration
CTAB	A cationic detergent that solubilises membranes and forms a complex with DNA allowing selective precipitation by lowering salt conc. <0.5 M, or adding isopropanol. May be useful for plants and matrices high in polysaccharides as these remain in solution (N.B. at higher salt concentrations contaminants are precipitated and DNA remains in solution).	1–2% (vol/vol) (will precipitate out at <15 °C)
Guanidine isothiocyanate	A chaotropic agent and strong protein denaturant when used at high salt conc. Inhibits nucleases.	4–5 M
Phenol	A protein denaturant. When added to crude aqueous DNA extracts, proteins collect in the organic phase or at the interface, DNA is maintained in the aqueous phase.	1:1 to aqueous solution
Sarkosyl	An anionic detergent used instead of SDS due to its higher solubility (SDS is insoluble in high salt conc.)	0.5–2.0% (vol/vol)
SDS (sodium dodecyl sulfate)	Anionic detergent and protein denaturant. Disassociates DNA-protein complexes. Foams easily when shaken.	0.5–2.0% (vol/vol)
Triton-X series	A series of gentle, non-ionic detergents that solubilise proteins without denaturation.	0.5% (vol/vol)
Tween series	A series of gentle, non-ionic detergents that solubilise proteins without denaturation.	0.5% (vol/vol)

agents. The concentration of enzyme used should therefore be empirically determined. Common enzymes used in cell and membrane lysis are listed in Table 4.3.

4.4.3.4 *Concentration of EDTA in Extraction Buffer*

The concentration of EDTA in the extraction buffer should be optimised to minimise the activity of endogenous DNases thereby maintaining the integrity

Table 4.3 Common enzymes used in cell and membrane lysis.

Enzyme	Description	Approx. conc.
Proteinase K	A non-specific protease, not inactivated by metal ions or chelating agents. Full activity over pH 6.5–9.5. Frees nucleic acids from adhering proteins. Activity stimulated by denaturing agents (SDS and urea).	0.1–0.2 mg ml^{-1}
Pronase	A mixture of proteases, can be a cheaper alternative to Proteinase K.	0.5–1 mg ml^{-1}
Lysozyme	Used with EDTA to break down cell wall or membranes in bacterial DNA extractions.	1–5 mg ml^{-1}
Zymolase/ Chitinase	Digests chitinous cell walls of fungi that may be resistant to mechanical forces.	1 mg ml^{-1}
Lyticase	Used for yeast cell wall degradation.	20 units ml^{-1}

of the DNA. EDTA acts by chelating the co-factor magnesium and other divalent cations. For calcium-rich samples EGTA, a specific complexing agent for calcium, can also be added to the extraction buffer.

4.4.3.5 Concentration of Salt in Extraction Buffer

Salt creates an isotonic environment to stabilise free nucleic acids (for example, phosphate buffered saline; 50 mM phosphate buffer, pH 7.4, 0.9% NaCl).

4.4.3.6 Extraction Buffer pH

The most common buffer used in molecular biology is Tris, with a pH range of 7.0–9.0. Other biological buffers are available which may be more suitable for some applications. For example, when working with acidic samples (such as certain foodstuffs) the pH may drop during the extraction procedure, therefore a buffer with a lower pH buffering range may be required. Extraction buffers should be carefully adjusted to the required pH. This should be carried out after all the components of the buffer are fully dissolved and, if required, the buffer has been autoclaved and cooled to room temperature. If stock extraction solutions are prepared, the pH of the diluted, final working solution should be checked and adjusted as necessary.

4.4.3.7 Excessive Damage of the DNA Analyte

Damage can be inflicted on DNA during the extraction process causing shearing and degradation. DNA molecules are susceptible to fracturing if heated, sonicated, ground in liquid nitrogen or forced through small cavities such as pipette tips to an excessive degree. Where these procedures are often essential in an extraction process, tolerance levels of such treatments should be determined empirically for a given sample. If high molecular weight DNA is required all such physical stress should be avoided, and gentler enzymatic lysis methods employed.

4.4.4 Separation of Nucleic Acids from Cell and Matrix Debris

Proteinaceous material is usually disrupted by proteinase K, denaturants and detergents in the extraction buffer. Deproteination and removal of debris has traditionally been achieved by organic extraction with phenol:chloroform and chloroform, where the DNA remains in the aqueous phase and all debris and proteins either collect in the organic phase or sediment at the interface. However, because of the toxicity of phenol many laboratories have switched to alternative methods such as commercial DNA affinity columns or magnetic bead-based chemistries, which can eliminate the need for dangerous organic solvents during extraction.

Factors affecting the validity of separation of nucleic acids have been identified as follows:

4.4.4.1 Phenol Quality

High-quality phenol should always be used which has been correctly buffered (pH 7–pH 8) and stored in the dark at 4 °C for no more than 1 month or at –20 °C for longer periods. 8-hydroxyquinoline (0.1%) stabilises phenol by retarding oxidation.

4.4.4.2 Inefficient Phenol Extraction and Removal

A sample may require more than one phenol extraction to remove potential contaminants. Care should be taken when removing the aqueous phase; the interface containing the debris should not be disturbed. It may be prudent to sacrifice a small volume of the aqueous phase close to the interface.

If phenol is used, traces should be removed from the sample by a chloroform:isoamyl alcohol (24:1) extraction to avoid inhibition of enzyme activity in downstream processes.

4.4.5 Additional Purification of DNA

Potential inhibitors of subsequent analyses can be removed at various stages of the extraction process. This is typically achieved in two ways. Firstly, reagents can be added to the extraction buffer that chelate or inactivate inhibitors. Secondly, an additional column-cleaning step can be included, which may either selectively bind DNA, allowing inhibitors to be eluted, or selectively bind inhibitors allowing the DNA to be eluted. When magnetic-beads are employed, additional wash steps may be incorporated to remove inhibitors.

Factors which can aid the additional purification of DNA have been identified as follows:

4.4.5.1 Composition of Extraction Buffer

Certain reagents such as those given in Table 4.4 can be added to the extraction buffer in order to chelate or inactivate inhibitors.

4.4.5.2 Column Cleaning

Crude DNA extracts can be cleaned by passing the extract through a resin column. These can work in different ways, either by binding the DNA and washing through contaminating agents or by binding the potential contaminant and allowing the purified DNA to pass through. Common column matrices available for purifying DNA are listed in Table 4.5.

4.4.5.3 RNase Treatment of the Sample

RNA contamination can interfere with some analytical procedures or quantification techniques. This may be particularly relevant when working with

Table 4.4 Reagent additives to extraction buffers that can remove potential inhibitors.

Additive	Reported action
PV(P)P 40 (polyvinyl(poly)pyrrolidine) (1–2% vol/vol)	Included in buffers for extraction of plants and soils rich in polyphenols. Assists in the adsorption of phenolics. Polyphenols can be oxidised by phenol oxidases into compounds that form complexes with nucleic acids causing damage to DNA and inhibit analyses involving enzymes. Should be prepared by acid wash.[6]
DTT (dithiothreitol) (1 mM)	Antioxidant, considered superior to β-mercaptoethanol as it is odourless and has less of a tendency to be oxidised by air.
DIECA (diethyl-dithiocarbamic acid) (4.0 mM/0.1%)	Phenol oxidase inhibitor. Inhibits the oxidation of polyphenols to quinonic substances that damage DNA.
Ascorbic acid (5 mM)	Strong reducing agent/antioxidant.
β-mercaptoethanol (0.2–5.0% vol/vol)	Antioxidant/reducing agent, protects sulfhydryl groups of enzymes against oxidation. Add to extraction buffer just before use.
Cysteine (10 mM)	Antioxidant.

Table 4.5 Common column matrices or bead-based chemistries available for purifying DNA (used according to the manufacturers' instructions).

Column/bead/matrix	Reported action
Hydroxyapatite	A form of calcium phosphate that binds dsDNA selectively in a mixture of nucleic acid types.
Gel filtration	Purifies by size fractionation. Examples include Sephadex -G 50-G 200 and CL6B Sepharose, which excludes smaller fragments (< 194 bp). Spin column formats are available.
Silica particles	Absorbs DNA, examples include glassmilk/glass fog. Qiagen surface modified silica gel acts as an anion exchange resin. A variety of silica-coated bead-based chemistries are also available.
Ultrafiltration	Membrane capture of DNA while smaller contaminants pass through. Examples include Centricon® concentrator (Amicon).
Anion exchange resins based on cellulose, dextran or agarose (sold under various commercial names)	Binds the strongly anionic DNA. The use of anionic detergents such as SDS should be avoided as these will also bind. Examples include DEAE Sephacel/Sepharose/Sephadex.
Cation exchange resins (Chelex® 100, Bio-Rad)	Chelating resin binds metals and other potential inhibitors, DNA should not be stored long term as inhibitors may be released over time. Yields partially single-stranded DNA therefore may bias quantification using intercalating dyes.
Paramagnetic resins	Magnetic bead-based resins or particles eliminate contamination and the need for high levels of dangerous solvents, for example the DNA IQ™ System from Promega or MagAttract® chemistry from QIAGEN.

RNA-rich tissues such as liver and kidney. Digestion of RNA can be undertaken at various stages of the extraction process if the buffer conditions are suitable. Care must be taken to inactivate DNases present in commercial RNase preparations according to the manufacturer's instructions. A stock solution of 10 mg ml^{-1} is normally prepared and used at a working concentration of approximately 100 μg ml^{-1}.

4.4.6 Precipitation and Concentration of DNA

This step in the procedure can be carried out for a number of reasons; firstly, to change the solvent, perhaps to a suitable buffer for storage purposes; secondly, to remove certain non-precipitated contaminants; and thirdly, to concentrate the DNA to a required working molarity. Optimisation of this stage in the extraction process is perhaps the most neglected step. Additional attention at this point could enhance both reproducibility and yield of extracted DNA. A review of the practical aspects of alcohol precipitation is given by Winfrey et al.[7]

Factors affecting the validity of DNA extraction due to the concentration or precipitation of DNA have been identified as follows:

4.4.6.1 Volume and Temperature of Alcohol Used and Precipitation Times

Either a 2× volume of ethanol or a 0.6× volume of isopropanol is usually recommended. DNA precipitations are undertaken at room temperature or at −20 °C. If the concentration of DNA is low, yields may be increased by overnight incubations at −20 °C, while for samples known to contain high quantities of DNA 10 minutes at room temperature may be sufficient. Centrifugation times and speed can also be increased for samples of low yield, however this may make the pellet more compact and harder to re-dissolve. Isopropanol is useful when smaller volumes are required, however it is less volatile and has more of a tendency to co-precipitate salts than ethanol.

4.4.6.2 Concentration and Type of Salt

In order for a nucleic acid precipitate to form, there must be at least 0.2 M concentration of a monovalent cation to shield the negative charge of the nucleic acid phosphate groups and allow aggregation of nucleic acid strands.[7] Therefore, the addition of salt to samples extracted with common buffers, which have comparatively low salt concentrations, may help the alcohol precipitation process if added to the final concentrations given in Table 4.6. However, if a high salt concentration buffer were used, the addition of further salts could reduce the yield of precipitated DNA. Washing the pellet with 70% ethanol can remove some of the precipitated salts.

4.4.6.3 Degraded DNA

When degraded or compromised DNA is expected from an extraction process (for example, when working with a highly processed sample or

Table 4.6 Recommended salt concentrations used in alcohol precipitations.

Salt	Stock concentration	Final concentration	*Specific use*
Sodium acetate	3.0 M pH 5.2	0.3 M	Generally used for DNA.
Ammonium acetate	7.5 M	2.5 M	Reduces co-precipitation of free dNTPs.
Sodium chloride	5 M	0.1 M	Used for samples containing SDS (SDS remains in solution).

paraffin-embedded sample), it should be remembered that short molecules of DNA (<200bps) are precipitated inefficiently by ethanol. Precipitation can be improved by the addition of $MgCl_2$ or glycogen to a final concentration of 10 mM and 10 µg ml^{-1} respectively.

4.4.6.4 DNA Concentration

If small-scale or low-yield extraction procedures are undertaken, DNA could be lost at the precipitation step. Addition of glycogen (to a final concentration of 10 µg ml^{-1}) can act as an efficient, inert co-precipitant of low concentrations of DNA. Commercial columns for concentrating DNA, such as Centricon® ultrafiltration tubes, may also be used as an alternative to alcohol precipitation. Correct choice of unit with the appropriate nucleotide cut off is required to avoid loss of degraded DNA. Commercial kits dedicated to extraction from samples with small amounts of DNA are also available, for example the QIAamp® DNA Micro Kit, where the conditions of the extraction and column formulation are optimised for low amounts of DNA.

4.4.6.5 Pellet Loss

Centrifugation tubes should be placed consistently in a centrifuge in such a way that the position of the DNA pellet is always known. In this way, even when the pellet is translucent and difficult to see, contact with the pellet, possibly causing it to dislodge, can be avoided.

4.4.6.6 Pellet Incompletely Re-suspended

High molecular weight DNA might take several hours to re-dissolve completely and pure DNA may dissolve more quickly than contaminated DNA pellets. Vigorous mixing or vortexing at this stage could damage the integrity of the DNA. In order to obtain maximum yield, integrity and reproducibility of DNA in the extraction process, pellets should be re-suspended at 4 °C overnight without agitation. If overnight incubation is insufficient, re-suspension can be undertaken by placing the sample at 45–50 °C for 15–20 minutes with occasional gentle inversion. It should also be remembered that up to 50% of the DNA may be smeared on the walls of the tube rather than being contained in the pellet.

4.4.6.7 Alcohol Precipitated Inhibitors

If it is found that inhibitors of certain analytical techniques are also precipitated, alternative methods of DNA concentration may be used, such as ultra-filtration columns (Table 4.5). For high-yield extractions, DNA can be removed by spooling or collecting the DNA precipitate on a sterile glass hook. Some of the salts or inhibitors that are precipitated by 95% ethanol may also be removed by a 70% ethanol wash of the pellet after precipitation.

4.5 Automation of DNA Extraction

Automated techniques are now commonplace in the laboratory. With downstream applications such as PCR and sequencing being handled in a high-throughput manner, the bottleneck has now, in many cases, become the extraction step. With the advancement of laboratory automation, there is a wealth of reliable instrumentation available to mechanise the early sample processing steps in the laboratory.

Many of the DNA extraction procedures discussed in this chapter are labour intensive. By automating the procedure, hands-on time can be drastically reduced. Furthermore, such automation can improve processing efficiency and reduce human errors, which multiply as the complexity of the manual extraction procedure increases.[8] In laboratories where DNA extraction from complex sample types is routine, such as those processing forensic crime scene samples, automation of the DNA extraction step frees up technician time for sample inspection or downstream STR analysis.[9] Elimination of errors, and thus repeat extractions, enables technicians to focus on other tasks and leads to an overall cost saving in the laboratory. Overall incorporation of extraction instrumentation into the laboratory standardises processing, decreases use of resources and thus reduces cost.

Regardless of the downstream application or instrument, all DNA isolation steps including sample preparation, cell or membrane lysis, protection and stabilisation of released DNA and separation of nucleic acids from cell debris or sample matrix must still be suitably accomplished on the robot. However, in the cases of DNA extraction from difficult sample types such as bones or teeth, additional offline up-front sample preparation and/or lysis steps may be performed.[10] Typically, a proteinase K digestion step or tissue disruption step using a mixer mill or grinder is performed before the analyst loads the sample on the extraction robot.

Instrumentation for DNA extraction is available in a wide range of throughput capacities. Robots are available to process as few as six samples, as well as equipment to rapidly process multiples of 96 well plates.[9,11,12] All chemistries and techniques discussed earlier in this chapter are amenable to automation, however many laboratories choose off-the-shelf commercial kits specially designed for robotic systems to decrease optimisation time. Many of these kits are validated on robotic systems and thus method-development time is virtually eliminated. Due to the popularity of automated extraction, protocols and references are available which highlight DNA isolation from even the most obscure sample types.

Instrument platforms are available in open and closed formats, the former allowing for the processing of multiple types of chemistries and programming manipulations by the analyst, while the latter typically is associated with particular chemistries with fully optimised and validated protocols contained in the software. Both formats have their advantages, and depending on the needs of the laboratory, one or the other is often preferred. Laboratories interested in their own frequent changes and manipulations to the protocols should choose open platforms. In these cases the technicians are able to make programming changes themselves. In instances where laboratories would like to standardise on ready validated protocols, closed systems are more suitable. In such cases, technician turnaround or inexperience does not tend to affect the overall success of the extraction process as changes cannot be made by the user to the software or protocol, and the automated procedure often requires only sample loading and pushing a few buttons.

The choice of instrument platform is also dependent on the needs of the laboratory. Specialised extraction robots are available which may also incorporate PCR set-up stations to further streamline the laboratory workflow. Additionally, liquid handling robots may be employed, which can accommodate DNA extraction procedures and PCR set-up as well as other liquid dispensing laboratory tasks.

Regardless of the instrument platform and chemistry, the analyst must ensure that the method performs within reasonable comparison to the manual methods currently used in the laboratory. Thus, even with validated, optimised commercial kits developed for instrumentation, a laboratory validation process is required to ensure yield, sensitivity and specificity are suitable to the laboratory's current standards.[13,14] Furthermore, just as with manual DNA extraction procedures, automated protocols and methods should be continually reviewed and validated to ensure proper quality control. Cross-contamination is always a risk during DNA extraction, and the appropriate controls must be incorporated to ensure that contamination has not occurred during the movement of mechanical elements of the instrument over open tubes or plates on the platform. However, the design of current robotic systems and software keeps cross-contamination issues to a minimum and when paired with the appropriate chemistry can achieve DNA isolation with high sensitivity and specificity.

4.6 DNA Extraction Protocols

This section provides the reader with some sources of well developed DNA extraction protocols (Table 4.7) and recent method developments. The

Table 4.7 DNA extraction methodologies. A range of commercial kits are available that have been developed and optimised for many of the listed sample types.

Application/matrix	Extraction method references
Plant material	
Fresh or hydrated plant tissue	CTAB[15]
Removal of contaminating proteins and polysaccharides	Hot SDS[15]
Material rich in polysaccharides and polyphenols	High potassium acetate[15] CTAB/PVP/chloroform[16]
Higher molecular weight DNA	Enzymic cell wall digestion[17]
High levels of secondary metabolites	Various[18]
Various plant tissues, herbarium samples and woody plants	CTAB/PVP/ascorbic acid/DIECA followed by chloroform/IAA[19]
Small scale and rapid methods for plants	Comparison of methods[20]
Various, including difficult plant species	Various[5]
Blood, tissue and body fluids	
Blood, tissue and cell culture	Various methods[21]
Biological evidence material (blood stains, hair, tissue, sperm cells and epithelial cells)	Various techniques[22,23] Chelex®[24]
Direct PCR in whole blood	AnyDirect buffer system[25]
Blood clots from serum separator tubes	Wire mesh centrifugation[26]
Cell free circulating nucleic acid	Additional centrifugation[27] Addition of formaldehyde[28] Automation[29]
Ancient bone tissue and hair samples	Comparison of three different techniques[30] Assessment of washing steps[31]
Animal and insect samples	
Bird feathers and pulp	Chelex®-100/proteinase K[32]; simple, non-hazardous[33]
Insects	CTAB/proteinase K[34]
Micro-organisms	
Range of bacteria and fungi	Review of techniques[35]
Fungal species	Various[5,36] Method comparison[37]
Soil microbes	Various[38,39] Comparison of methods[40,41]
Microbial pathogens from cheeses, milk and meats	Various[42]
Microbes from clinical samples including gastric biopsy, clinical swabs, nasopharyngeal samples, saliva, stool and blood samples	Various[43] Method comparisons[44,45] Automated, magnetic bead based[46]
Other samples	
Nucleic acids from aquatic environments	Various[39,47]
Human DNA from cigarette butts, postage stamps and other saliva stained material	SDS/Proteinase K[48]
Biological fluids from cotton swabs	Cellulase digestion[49]

references provide a range of methods and links to further publications, and are provided as a source of useful methodological details, performance comparisons and valuable troubleshooting advice.

4.7 Summary

Traditional DNA extraction methodologies employing chemicals such as SDS, proteinase K and phenol are now reasonably well established.[50] These methods tend, however, to be time consuming and involve multiple liquid transfer operations. Alternative commercial kits often offer reduced hands-on time and cleaner approaches to extractions; however, they can be more expensive and limited to very specific applications. In an analytical environment neither of the approaches mentioned may be ideal and the situation can be further complicated by the sample matrix composition. Whilst simple approaches may be preferable, complex matrices and non-ideal samples may demand additional clean-up procedures. Measurement of DNA yield in itself is not sufficient to determine the suitability of an extraction methodology. The quality, encompassing purity and integrity, of the DNA analyte can also be important in ensuring suitability for downstream measurements. An inappropriate choice or a sub-optimal extraction methodology could have significant consequences for subsequent analyses, which may have to be repeated, or produce false negative results. Automation of the entire extraction procedure increases throughput and reduces analyst errors, often resulting in a more accurate and cost-effective DNA isolation step in the laboratory workflow. Validation of sampling procedures, sample storage, sample preparation and DNA extraction should all be considered vital to the production of quality data in subsequent analyses whether these are accomplished manually or on automated platforms.

References

1. N. B. Siddappa, A. Avinash, M. Venkatramanan and U. Ranga, *Biotechniques*, 2007, **42**, 188.
2. J. Day and G. Stacey, *Cryopreservation and Freeze-drying Protocols (2 Rev ed., Methods in Molecular Biology Series)*, Humana Press Inc., Totowa, NJ, 2006, ISBN13 978-1-58829-377-0.
3. RSC, *General Principles of Good Sampling Practice*, The Royal Society of Chemistry, Cambridge, 1995, ISSN 0-85404-412-4.
4. L. Kokilera, *Comp. Biochem. Physiol.*, 1995, **111B**, 35.
5. K. Weising, H. Nybom, K. Wolff and W. Meyer, *DNA Fingerprinting in Plants and Fungi*, CRC Press, Boca Raton, FL, 1994, ISBN 0-84938-920-8.
6. http://www.emdbiosciences.com/SharedImages/TechnicalLiterature/1_Detbooklet_2001.pdf (accessed July 2007).

7. M. R. Winfrey, M. A. Rott and A. T. Wortman, *Unraveling DNA: Molecular Biology for the Laboratory,* Benjamin Cummins, Menlo Park, CA, 1997, ISBN-13 978-0-132-700344.
8. C. A. Crouse, S. Yeung, S. Greenspoon, A. McGuckian, J. Sikorsky, J. Ban and R. Mathies, *Croat. Med. J.*, 2005, **46**, 563.
9. S. A. Montpetit, I. T. Fitch and P. T. O'Donnell, *J. Forensic Sci.*, 2005, **50**, 555.
10. http://www1.qiagen.com/literature/handbooks/PDF/GenomicDNAStabilizationAndPurification/FromClinicalSamples/AN_BREZ1/1031410_0206_AN_EZ1_bones_lr.pdf (accessed July 2007).
11. S. A. Greenspoon, J. D. Ban, K. Sykes, E. J. Ballard, S. S. Edler, M. Baisden and B. L. Covington, *J. Forensic Sci.*, 2004, **49**, 29.
12. M. Rechsteiner, *Forensic Magazine*, 2006, **(April–May)**, 9.
13. E. Gobbers, T. A. Oosterlaken, M. J. van Bussel, R. Melsert, A. C. Kroes and E. C. Claas, *J. Clin. Microbiol.*, 2001, **39**, 4339.
14. M. Nagy, P. Otremba, C. Kruger, S. Bergner-Greiner, P. Anders, B. Henske, M. Prinz and L. Roewer, *Forensic Sci. Int.*, 2005, **152**, 13.
15. S. Wilke, in *Plant Molecular Biology: A Laboratory Manual*, ed. M. S. Clark, Springer, Berlin, 1997, p. 3, ISBN-13 978-3-540-584056.
16. P. Towner, in *Essential Molecular Biology: Vol 1 (Practical Approach)*, ed. T. A. Brown, Oxford University Press, NY, 2000, p. 47. ISBN-13 978-0-199-636426.
17. J. F. Manen, O. Sinitsyna, L. Aeschbach, A. V. Markov and A. Sinitsyn, *BMC Plant Biol.*, 2005, **5**, 23.
18. E. A. Friar, *Methods Enzymol.*, 2005, **395**, 3.
19. C. Neal Stewart, Jr., in *Fingerprinting Methods Based on Arbitrarily Primed PCR: Random Amplification Methods*, ed. M. R. Micheli and R. Bova, Springer, Berlin, 1997, p. 25, ISBN-13 978-3-540-612292.
20. H. J. Rogers, N. A. Burns and H. C. Parkes, *Plant Molecular Biology Reporter*, 1996, **14**, 170.
21. E. D'Ambrosio and E. Pascale, in *Fingerprinting Methods Based on Arbitrarily Primed PCR: Random Amplification Methods*, ed. M. R. Micheli and R. Bova, Springer, Berlin, 1997, p. 15, ISBN-13 978-3-540-612292.
22. G. F. Sensabaugh, in *Ancient DNA*, ed. B. Herrman and S. Hummel, Springer, New York, 1996, p. 141, ISBN-13 978-0-387-943084.
23. S. Easteal, N. McLeod and K. Reed, *DNA Profiling. Principles, Pitfalls and Potential*, Harwood Academic Publishers, Switzerland, 1991, ISBN-13 978-3-718-651900.
24. J. M. Willard, D. A. Lee and M. M. Holland, in *Forensic DNA Profiling Protocols (Methods in Molecular Biology)*, ed. P. J. Lincoln and J. Thomsom, Humana Press Inc., Totowa, NJ, 1998, p. 9, ISBN-13 978-0-896-034433.
25. Y. G. Yang, J. Y. Kim, Y. H. Song and D. S. Kim, *Clin. Chim. Acta.*, 2007, **380**, 112.
26. S. Se Fum Wong, J. J. Kuei, N. Prasad, E. Agonafer, G. A. Mendoza, T. J. Pemberton and P. I. Patel, *Clin. Chem. (Washington, DC, US)*, 2007, **53**, 522.

27. K. Page, T. Powles, M. J. Slade, M. T. De Bella, R. A. Walker, R. C. Coombes and J. A. Shaw, *Ann. N. Y. Acad. Sci.*, 2006, **1075**, 313.
28. R. Dhallan, X. Guo, S. Emche, M. Damewood, P. Bayliss, M. Cronin, J. Barry, J. Betz, K. Franz, K. Gold, B. Vallecillo and J. Varney, *Lancet*, 2007, **369**, 474.
29. D. J. Huang, B. G. Zimmermann, W. Holzgreve and S. Hahn, *Ann. N. Y. Acad. Sci.*, 2006, **1075**, 308.
30. H. Nielsen, J. Engberg and I. Thuesen, in *Ancient DNA*, ed. B. Herrman and S. Hummel, Springer, New York, 1996, p. 122, ISBN-13 978-0-387-943084.
31. S. Amory, C. Keyser, E. Crubezy and B. Ludes, *Forensic Sci. Int.*, 2007, **166**, 218.
32. H. Ellergen, in *Ancient DNA*, ed. B. Herrman and S. Hummel, Springer, New York, 1996, p. 211, ISBN-13 978-0-387-943084.
33. S. M. Bailes, J. J. Devers, J. D. Kirby and D. D. Rhoads, *Poult. Sci.*, 2007, **86**, 102.
34. G. J. Hunt, in *Fingerprinting Methods Based on Arbitrarily Primed PCR: Random Amplification Methods*, ed. M. R. Micheli and R. Bova, Springer, Berlin, 1997, p. 29, ISBN-13 978-3-540-612292.
35. D. Park, *Methods Mol. Biol.*, 2007, **353**, 3.
36. G. C. Graham and R. J. Henry, in *Fingerprinting Methods Based on Arbitrarily Primed PCR: Random Amplification Methods*, ed. M. R. Micheli and R. Bova, Springer, Berlin, 1997, p. 21, ISBN-13 978-3-540-612292.
37. D. N. Fredricks, C. Smith and A. Meier, *J. Clin. Microbiol.*, 2005 **43**, 5122.
38. A. Saano, E. Tas, S. Pippola, K. Lindstrom and J. D. van Elsas, in *Nucleic Acids in the Environment*, ed. J. T. Trevors and J. D. van Elsas, Springer, Berlin, 1995, p. 49, ISBN-13 978-0-387-580692.
39. A. K. Bej and M. H. Mahbubani, in *PCR Technology; Current Innovations*, ed. H. G. Griffin and A. M. Griffin, CRC Press, Boca Raton, FL, 1994, p. 327, ISBN 0-849-386748.
40. Z. H. H. Yang, Y. Xiao, G. M. Zeng, Z. H. Y. Xu and Y. S. H. Liu, *Appl. Microbiol. Biotechnol.*, 2007, **74**, 918.
41. C. A. Whitehouse and H. E. Hottel, *Mol. Cell Probes.*, 2007, **21**, 92.
42. D. D. Jones and A. K. Bej, in *PCR Technology: Current Innovations*, ed. H. G. Griffin and A. M. Griffin, CRC Press, Boca Raton, FL, 1994, p. 341, ISBN 0-849-386748.
43. M. H. Mahbubani and A. K. Bej, in *PCR Technology: Current Innovations*, ed. H. G. Griffin and A. M. Griffin, CRC Press, Boca Raton, FL, 1994, p. 307, ISBN 0-849-386748.
44. A. Saukkoriipi, T. Kaijalainen, L. Kuisma, A. Ojala and M. Leinonen, *Mol. Diagn.*, 2003, **7**, 9.
45. K. Smith, M. A. Diggle and S. C. Clarke, *J. Clin. Microbiol.*, 2003, **41**, 2440.
46. M. Stormer, K. Kleesiek and J. Dreier, *Clin. Chem. (Washington, DC, US)*, 2007, **53**, 104.

47. J. H. Paul and S. L. Pichard, in *Nucleic Acids in the Environment*, ed. J. T. Trevors and J. D. van Elsas, Springer, Berlin, 1995, p. 153, ISBN-13 978-0-387-580692.
48. M. N. Hochmeister, O. Rudin and E. Anbach, in *Forensic DNA Profiling Protocols (Methods in Molecular Biology)*, eds. P. J. Lincoln and J. Thomson, Humana Press, Totowa, NJ, 1998, p. 27, ISBN-13 978-0-896-034433.
49. J. C. Voorhees, J. P. Ferrance and J. P. Landers, *J. Forensic Sci.*, 2006, **51**, 574.
50. J. Sambrook, E. F. Frisch and, T. Maniatis (ed.), *Molecular Cloning – A Laboratory Manual*, Cold Spring Harbor Laboratory Press, NY, 1997, ISBN-13 978-0-879-693091.

CHAPTER 5
DNA Quantification

PAUL A. HEATON AND JACQUIE T. KEER

LGC, Queens Road, Teddington, TW11 0LY

5.1 Introduction

The ability to quantify nucleic acids is central to many applications using DNA measurements. Current methods such as UV absorbance and fluorescence spectroscopy are used for many routine molecular biological applications. However comparability of results between laboratories and methods can be poor,[1,2] and inherent difficulties and constraints in the methods can introduce inaccuracy and uncertainty into the measurement process.

Quantification of total DNA may be required in many situations, including:

- The control of the amount of starting material for downstream analytical procedures following initial sample preparation;
- Clinical applications such as determining the efficacy of treatment in reducing bacterial or viral load, or monitoring disease progression in cancer patients;
- Determination of product regulatory compliance, such as Genetically Modified Organism (GMO) content of foods or the residual DNA content of biopharmaceuticals;
- To determine the yield obtained from DNA extraction methods for comparative or validation purposes.

For applications such as the characterisation of the GMO content of foodstuffs, where legislative and regulatory limits must be observed, accuracy of quantitative analysis is critical.[3] Development of accurate techniques for quantification and the availability of accurately quantified DNA standards and quantitative reference materials are routes to increasing the comparability of measurements between laboratories. In this chapter, both current methods widely used for DNA quantification and emerging methods with the potential to provide greater quantitative accuracy will be described, and potential problems with available methods will be highlighted.

5.2 Measurement of DNA Concentration Using Ultraviolet Spectroscopy

One of the most commonly used and simplest techniques for DNA quantification is spectroscopic determination by UV absorption. All nucleic acids absorb strongly in the UV region due to the heterocyclic ring structures associated with each of the four bases. Typically maximal absorption is observed at a wavelength of around 260 nm, although this is pH dependent.

Measurement of DNA concentration using UV absorbance is a relatively simple process, involving the following steps:

(i) Dilution of the DNA solution in either a buffer or double distilled water to a concentration giving an A_{260} of 1 optical density (OD) unit or less;
(ii) Determination of the absorbance value of the solution, taking into account the background absorbance of the diluent and any dilution factor applied;
(iii) Conversion of the adjusted absorbance value into a concentration value for the solution.

The determined absorbance values are converted to concentrations using the Beer-Lambert Law (Equation 5.1), which states that the measured absorbance (A) of a DNA solution is determined by the DNA concentration (c), the path length (l) and the extinction coefficient (ε). The Beer-Lambert law predicts a linear change in absorbance with concentration, although the linear relationship does break down at high DNA concentrations (with readings higher than 1.5 OD units it is advisable to dilute and re-measure the sample).

$$A = \varepsilon c l \quad (5.1)$$

The path length is usually fixed at 1 cm by the spectrophotometer, although one of the newer instruments on the market has been designed to work with much lower sample volumes and has a path length varying between 0.2 and 1 mm. The value of the extinction coefficient (ε) may be calculated, determined experimentally or, more usually, approximated. Determination of sample DNA concentration using UV spectroscopy can also be estimated by comparison to DNA quantification standards through preparation of a calibration graph.

5.2.1 Determining the Extinction Coefficient ε

Calculation of ε: The value of ε for a particular DNA is dependent on length and nucleotide sequence, the wavelength at which it is measured (usually 260 nm) and the pH of the solution. Each component base in the DNA contributes to ε, but the contribution is not simply additive since neighbouring bases interact with each other, reducing their extinction coefficients at 260 nm. This effect is known as the hypochromic shift. The value of ε can be calculated for short oligonucleotides although the complexity of genomic DNA requires a different approach.

Experimental Determination of ε: The most common method is to compare the absorbance of a given DNA sample before (A_0) and after (A_∞) total enzymatic digestion, for example with snake venom phosphodiesterase, to its component 5′-mononucleotides.[4] Assuming complete digestion, the ratio of $A_0:A_\infty$ can then be used in conjunction with the sum of the extinction coefficients of each of the separate bases (ε_{sum}) to calculate ε for the intact DNA under test (Equation 5.2). Again, the size and sequence of the DNA is required.

$$\varepsilon_{intact} = A_0/A_\infty (\varepsilon_{sum}) \qquad (5.2)$$

Approximation of ε: Usually, the length and sequence of an extracted genomic DNA is not available, which precludes the accurate determination of the extinction coefficient. In these situations, suitable approximations need to be made, which introduces uncertainty into the process. Typical values for ε per base are 6600 $M^{-1}\,cm^{-1}$ for dsDNA and dsRNA and 8500 $M^{-1}\,cm^{-1}$ for single-stranded nucleic acids.[5] The established relationships of 50 μg ml^{-1} dsDNA and 40 μg ml^{-1} ssDNA having an absorbance reading of 1 are derived from these values of ε.[6]

5.2.2 Practical Aspects of Measuring DNA Concentrations by UV Spectroscopy

Despite the apparent simplicity of UV determination of DNA concentration, practically it can be fraught with problems. As described above, the extinction coefficient used will influence the results obtained, and a range of other factors may also affect the determined DNA concentration.

5.2.2.1 Calibration of the Spectrophotometer

Calibration should be performed regularly, either in-house or by external contractors, using standard filters that are transparent to only certain, defined wavelengths. Low $A_{260}:A_{280}$ ratio values can, in some cases, indicate a problem with the spectrophotometer.[7] A number of materials are commercially available that may be used to check the performance of the spectrophotometer, which vary in quality and price. Several of the materials are traceable to National Institute of Standards and Technology (NIST) standards, such as the In-Spec® UV standards and potassium dichromate filled sealed quartz cells from Starna Cells, Inc. (Atascadero, CA, USA).

5.2.2.2 Cuvettes

These should be of good quality and transparent in the UV region. Care should be taken to handle cuvettes only on their non-optical surfaces, and they should be prepared for use by rinsing with 95% ethanol followed by double-distilled or ultra pure water and wiped dry with lint-free paper tissue. To avoid measurement bias through the use of mis-matched cuvettes, the absorbancies of cuvettes filled with blank solutions may be read and compared. If a significant and consistent difference is obtained, appropriate adjustment of sample readings can be made.

5.2.2.3 Sample Preparation

Care should be taken to ensure that DNA in the samples under test is fully dissolved before concentrations are measured. In addition, any particulate matter should be removed by centrifugation or further purifying the sample.

5.2.2.4 Reference Blank

This is a cuvette containing an identical solution to the cuvette that is being assayed, except the DNA component is excluded. A blank should always be used when measuring DNA concentrations to correct for any absorbance of the diluent containing the DNA.

5.2.2.5 Sample Dilution

Absorbance (A), and therefore concentration, is related to the logarithm of transmittance (T) of the sample, which is defined as the ratio of light passing through the sample (P_{sample}) as compared with the blank (P_{blank}) (Equation 5.3).

As a result of this relationship, errors in the measured transmittance at low concentrations have a far larger effect on the concentration measurement than those measured at higher DNA concentrations.

$$A = -\log T, \text{ where } T = P_{sample}/P_{blank} \quad (5.3)$$

Thus, when determining DNA concentrations using the spectrophotometer, a higher DNA concentration is preferable so long as the absorbance readings are in the linear range (usually A_{260} values of <1). Below an absorbance of 0.01 OD units measurements are not generally considered to be reliable, corresponding to a lower DNA concentration of approximately 0.5 µg ml^{-1}, although specifications may vary between instruments.

5.2.2.6 Light Source

The deuterium lamp in the spectrophotometer should be allowed to warm up before use (according to the manufacturers instructions) to allow the emissions from the source to stabilise. In addition, the bulb should be replaced when necessary.

5.2.2.7 Presence of Contaminants

Lack of specificity is a major drawback of the UV absorbance method. Any contaminant present in the sample that absorbs at 260 nm will contribute to the final DNA concentration, potentially leading to overestimation of the actual DNA concentration. Common contaminants of DNA that absorb at this wavelength range include RNA, proteins, EDTA and phenol. The ratio of measurements at $A_{260}:A_{280}$ can be used as a primary indicator of purity, with pure DNA characteristically exhibiting an $A_{260}:A_{280}$ absorbance ratio of 1.86. Lower values typically indicate co-purification of contaminants such as phenol or protein. The $A_{260}:A_{230}$ ratio may be used as an additional indicator, with acceptable measurements for nucleic acids in the range of 1.8–2.2 and much lower values typically indicating the presence of contaminants. Measurements at 320 nm may be used to detect the presence of particulates, as values of $A_{320}:A_{260}$ ratio higher than 0.1 indicate particulates or other undesirable materials in the preparation. Additionally, nucleic acid solutions used for absorbance measurements should ideally contain less than 1 mM EDTA, which absorbs strongly at the commonly used test wavelengths.

RNA

The similarity in the absorbance spectra of DNA and RNA make contamination difficult to detect. Although high levels of RNA may be present in DNA preparations, such contamination can be removed effectively by digestion with RNase A followed by a further purification step to remove the enzyme. An $A_{260}:A_{280}$ ratio of 2.0 is characteristic of pure RNA.

Proteins

Proteins are more easily distinguishable from DNA by absorbance spectrum, with a maximum of 280 nm with some absorbance at 260 nm. A protein concentration of $0.3\,\text{mg\,ml}^{-1}$ has an absorbance of approximately 0.01 OD at 260 nm and this value can be taken as an upper limit before an effect on the determined DNA concentration is observed. A simple but insensitive method to detect a protein contamination in a sample is to determine the $A_{260}:A_{280}$ ratio, as contamination with protein is indicated by values significantly lower than 1.8. A value of 0.6 is characteristic of pure protein. Care must be taken when using the $A_{260}:A_{280}$ ratio as an assessment of sample purity as the ratio is also dependent on the pH and ionic strength of the sample.[8] As pH increases, the absorbance at 280 nm decreases but the absorbance at 260 nm is unaffected; this results in the ratio having elevated values at higher pH. In contrast, increasing the ionic strength, for instance, the salt concentration tends to decrease both absorbancies but has the overall effect of increasing the $A_{260}:A_{280}$ ratio. In addition, as many proteins do not contain a high amount of aromatic amino acids, the $A_{234}:A_{260}$ ratio can also be used to check for purity; nucleic acids have an absorbance minimum at 234 nm and protein contamination is detectable as an increase in this ratio above 0.5. Protein contamination can be removed from the nucleic acid sample by digestion with Proteinase K followed by a further purification step.

Phenol

Phenol, like proteins, has an absorbance spectrum that is readily distinguishable from nucleic acids; it has an absorbance maximum at 264 nm and also absorbs strongly at 260 and 280 nm. An approximate 1:100,000 dilution of tris-saturated phenol has an A_{260} of 0.01 OD and this can be considered a lower detection limit. As with proteins, the $A_{260}:A_{280}$ ratio can be used to indicate a phenol contamination problem. Any residual phenol may be removed by a simple chloroform extraction.

5.3 Determination of DNA Concentration by Fluorescence Spectroscopy

Determination of DNA concentration by fluorescence spectroscopy relies on the fluorescent enhancement of a dye on intercalation into a nucleic acid. A range of dyes is now available with improved sensitivities, some of which bind specifically to a particular type of nucleic acid, such as double- or single-stranded DNA and RNA. Measurements made with such dyes are thus free from interference from other nucleic acid contaminants, which can give erroneous measurements when using UV spectroscopy. Concentration measurements are made as follows:

1. The DNA is diluted to a suitable concentration in ultra pure water or a suitable buffer, such as Tris-EDTA (TE), so that on addition of the fluorescent dye the fluorescence is within the measurement range of the instrument;
2. A fluorescent dye is added, which should be at a concentration sufficient to saturate all possible intercalation sites on the DNA analyte and should be kept at a fixed concentration for the series of measurements;
3. A fluorescence measurement is made of the unknown sample, and a calibrant at a range of concentrations, under the same conditions;
4. The fluorescence value is then converted into a DNA concentration using an empirically determined conversion factor.

5.3.1 Preparation of a Calibration Graph

The fluorescent conversion factor can be determined by means of a calibration graph, on which the fluorescence values from a dilution series of DNA standards has been plotted against known concentrations (Figure 5.1). The variation in fluorescence with concentration should be linear provided the DNA concentration is not excessively high and sufficient dye has been used to saturate all possible intercalation sites. It is important to ensure that the

DNA Quantification

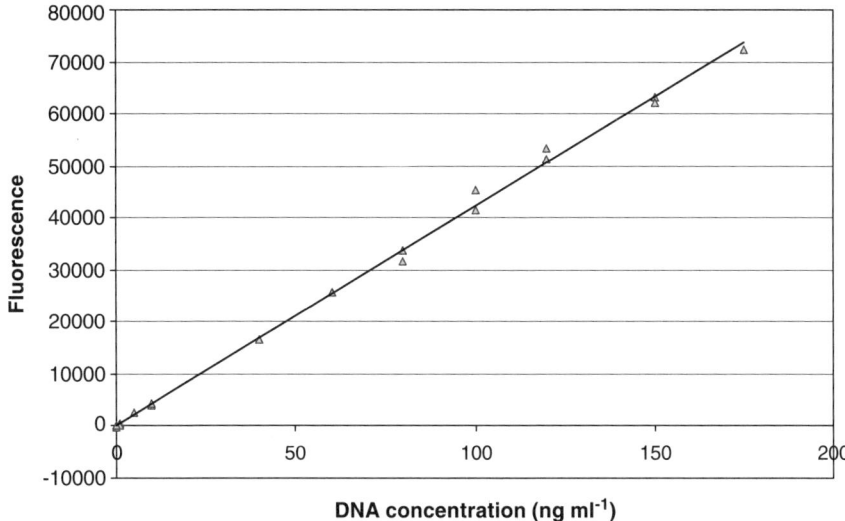

Figure 5.1 Calibration curve for PicoGreen® determination of DNA concentration, showing the fluorescence readings of the standard dilution series, and the correlation coefficient.

concentration ranges of the standards give fluorescence values that are within the linear dynamic range of the fluorometer or fluorescence plate reader being used for the assay. The gradient of the calibration graph can then be used as a conversion factor to determine sample DNA concentrations.

Concentration values of the standards could have been measured by another established methodology such as UV spectroscopy, or supplied by a manufacturer. The accuracy of the quantification process is significantly influenced by the quality of the standards used, and thus consistency in the source and use of standards is central to acceptable method performance. Further, as determination of absolute DNA quantities remains technically challenging, then the accuracy of any assigned standard values must be critically assessed to determine the reliability of standards used for fluorescent spectroscopy.

5.3.2 Practical Aspects of Measuring DNA Concentrations by Fluorescence Spectroscopy

A range of factors may influence the accuracy and reliability of the quantitative analysis, and these should be considered when determining DNA concentrations by fluorescence measurements.

5.3.2.1 Sample Preparation

The DNA sample should be fully dissolved and free from particulate material before measurements are made.

5.3.2.2 Reference Blank

Reference blanks should be used that have the same composition as the test samples, except with the exclusion of the DNA component.

5.3.2.3 DNA Standard

The fluorescent intensity of a given DNA, after intercalation of a fluorescent dye, is dependent on the length and, in some instances, the sequence of the DNA used. Ideally, the DNA standard used for the serial dilution should be pure target DNA but this is not always feasible. In such circumstances, a compromise needs to be made and a commercially available pure DNA (for example ultra pure calf thymus DNA) of similar length may be used. In this case, it is assumed that a similar number of dye molecules intercalate into the selected standard DNA as into the target, for instance, the dye does not display any sequence specificity.

5.3.2.4 Selecting the Dye

A dye should be selected that is sensitive enough to detect DNA concentrations in the range anticipated in the test samples and if possible be selective for the nucleic acid of interest.

5.3.2.5 Dye Concentration

The amount of dye used should be constant in the test samples and the DNA standard used to construct the calibration curve. This amount should be sufficient to saturate the range of DNA quantities under test. The dye saturation point may be determined empirically by measuring the fluorescence of samples containing a fixed amount of dye and a varying amount of DNA. The point at which the plot deviates from linearity is the point at which DNA saturation has been exceeded.

5.3.2.6 Microtitre Plates

When using a microtitre format and plate reader to measure fluorescence, good-quality optical plates should be chosen. In addition, black or opaque walled plates are recommended to minimise interference from adjacent wells when measurements are being made.

5.3.2.7 Measurement Conditions

The conditions under which the standard DNA is prepared and measured must be identical to those used for the sample DNA. Important factors include final concentration of dye, the buffer used to dilute the samples, final salt concentrations and the fluorometer settings. The range of standard concentrations used should span the expected range of sample concentrations, with the majority of the samples falling in the mid-range.

5.3.3 Fluorescent Dyes

A wide range of fluorescent dyes is commercially available. The choice of dye can affect the performance of the quantification assay, and it is therefore important to use an appropriate reagent for the level and type of DNA under test. Dye characteristics such as the excitation and emission spectra determine the compatibility with instrumentation. In addition, the required sensitivity of the assay needs consideration, with several physical properties such as the extinction coefficient, fluorescence quantum efficiency (the efficiency with which absorbed light is converted to emitted light) and the stability of the dye determining the fluorescence output of the dye–DNA complex. The labelling efficiency and effect of the dye on the physical properties of the DNA (such as electrophoretic mobility) may also need to be taken into account. Some of the most commonly used dyes, ethidium bromide, PicoGreen®, Hoechst 33258 and SYBR Green®, are discussed in more detail below; Table 5.1 summarises several relevant dye characteristics.

5.3.3.1 Ethidium Bromide

Although ethidium bromide is not usually employed with spectroscopic techniques, it is the traditional reagent for the fluorescent labelling and visualisation of nucleic acids. It has a broad specificity in that it will label both DNA and RNA. However it has a limited sensitivity, with a detection limit of approximately $100\,\text{ng}\,\text{ml}^{-1}$, since it has only a moderate affinity for nucleic acids and fluorescent enhancement on binding. Using ethidium bromide, samples can be stained after or during agarose gel electrophoresis, and the resultant fluorescence (induced by UV light excitation) is proportional to the amount of DNA. Using gel electrophoresis, comparison to a set of standards of known concentrations (usually lambda DNA or molecular weight markers) gives an estimation of DNA quantity in $\text{ng}\,\mu\text{l}^{-1}$. Although more time consuming than spectroscopy measurements, this approach is included as it is commonly used. Gel-based quantification also has the advantage of allowing simultaneous

Table 5.1 Characteristics of commonly used fluorophores.

Dye	Excitation maximum (nm)	Emission maximum (nm)	Quantum efficiency (QE)	Fluorescent enhancement on binding dsDNA
Ethidium bromide	518	605	0.15	~30 x
PicoGreen®	502	523	0.5	>1000 x
SYBR Green® I	497	520	0.8	>300 x
SYBR Gold	300 and 495	537	0.6	~1000 x
Hoechst 33258	352	461	0.42	~30 x
Propidium iodide	530	625	0.16	>10 x

semi-quantification of genomic DNA, estimation of RNA contamination and evaluation of DNA integrity.

Ethidium bromide is a known mutagen, so suitable precautions must be taken when handling this material. Ethidium bromide must also be disposed of appropriately, and methods for dealing with solutions containing ethidium bromide include chemical decontamination[9] and charcoal filtering.[10] Because of the potential hazards associated with ethidium bromide a range of safer dyes is now commercially available.[11]

5.3.3.2 PicoGreen®

PicoGreen® is a fluorescent dye that binds very specifically to double-stranded DNA but is little affected by single-stranded DNA, RNA and protein contaminants.[12] Since PicoGreen® has a very high affinity for double-stranded DNA and shows a high fluorescent enhancement on binding, the dye is sensitive and can be used across a wide range of DNA concentrations, typically from $25\,\text{pg}\,\text{ml}^{-1}$ to $1\,\mu\text{g}\,\text{ml}^{-1}$.[12]

When using PicoGreen®, care must be taken to control the conditions under which fluorescence measurements are made, as some contaminants may affect the level of fluorescent signal produced. Divalent metal ions, such as Mg^{2+}, Ca^{2+} and Zn^{2+}, have been shown to have a strong quenching effect on PicoGreen® fluorescence, although by contrast monovalent ions such as Na^+ and K^+ have only a modest effect on fluorescence and thus may be tolerated in quite high concentrations.[12] A variety of other reagents including sodium dodecyl sulfate (SDS) and phenol also have a detrimental effect on fluorescence. To ensure reliable concentration determination the test sample should be purified to remove such contaminants; the manufacturer's website is a useful source of information about the effects of a range of substances on PicoGreen® fluorescence.

5.3.3.3 SYBR Dyes

A range of newer cyanine dyes has been developed for gel staining and nucleic acid quantification, including the commonly used SYBR Gold and SYBR Green® I. These dyes provide a very high level of sensitivity for gel analysis, with as little as 25 pg and 60 pg bands being detectable, respectively. A further advantage is that the SYBR dyes are significantly less mutagenic than ethidium bromide, with a specific SYBR Safe reagent available for direct replacement of ethidium bromide for gel staining. In addition another cyanine dye has been developed, SYBR Green® II, which exhibits a larger fluorescence quantum yield when bound to RNA (~ 0.54) than to dsDNA (~ 0.36) and thus may be used for RNA and ssDNA quantification and gel visualisation.

5.3.3.4 Hoechst 33258

Hoechst 33258 similarly exhibits fluorescent enhancement on intercalation with DNA, and can bind to the minor groove of double-stranded DNA. Hoechst

33258 shows similar specificity to PicoGreen® for binding double-stranded DNA (and to a lesser extent single-stranded DNA) in preference to RNA, and is also unaffected by the presence of proteins. Again, conditions under which fluorescence measurements are made are important. Divalent metal ions such as Mg^{2+} (>10 mM) should be avoided as they reduce the fluorescent enhancement observed, whilst monovalent metal ions have very little effect on fluorescence.[13] A disadvantage of Hoechst 33258 is that it binds specifically to AT regions.[14] This can cause problems when the DNA used to construct the calibration curve has a different nucleotide composition from the test samples, and to ensure reliable results it is advisable to use suitably matched DNA in the analysis.

5.4 Quantification Using the Polymerase Chain Reaction

The advent of PCR during the mid-1980s[15] enabled measurements of target DNA across a wide dynamic range, and is sensitive to as little as a few copies. Conventional PCR analysis involves end-point detection of the products formed, and as the rate of product generation is not linear over the course of the reaction then extrapolating from the final amount of product to the initial amount of starting material is not straightforward. However, by defining the sensitivity of the reaction or by the use of reference samples, comparative standards or competitive mimics it is possible to generate data ranging through qualitative to semi-quantitative and quantitative. Such strategies all involve the detection or quantification of PCR products, using a range of methodologies including agarose gel electrophoresis in the presence of ethidium bromide, capillary electrophoresis and mass spectrometry. Detection methods vary in the accuracy and precision of analysis.

More recently the development of quantitative real time PCR[16] has enabled highly accurate quantification using appropriate calibration standards, and this will be detailed in Chapter 7.

5.5 Enzymatic Quantification of DNA

An alternative approach to DNA quantification has been developed, utilising energy from nucleotide triphosphates to generate light using a series of coupled enzymatic reactions. The method has been applied to the quantification of human DNA sequences by Promega for forensic purposes,[17] and is commercially available. The system uses two incubation steps to determine the amount of human DNA present in a sample (Figure 5.2). In the first stage a coupled enzymatic reaction takes place. The first reaction involves the addition of a pyrophosphate across the 3'-terminal bond of double-stranded DNA, resulting in the generation of a dNTP from the molecule. The dNTP produced then transfers the terminal phosphate to ADP present in the reaction. In the second incubation stage, the ATP is used by luciferase, resulting in the generation of

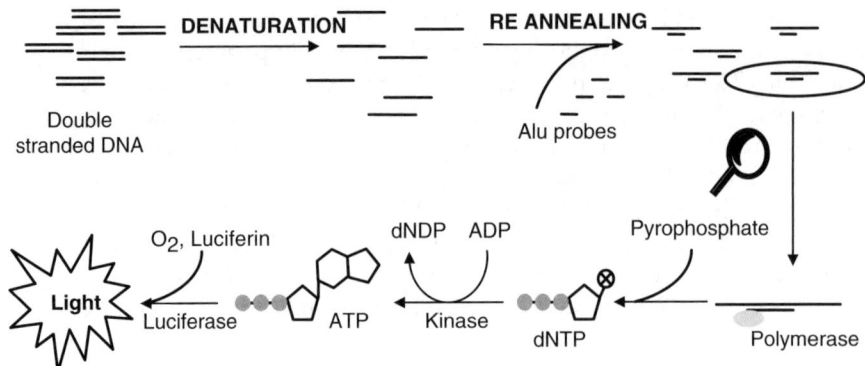

Figure 5.2 Schematic diagram illustrating the mechanism of enzymic DNA quantification utilising luciferase. In the example illustrated here, Alu probes directed towards repeat sequences in the human genome are used to enable specific quantification of human DNA in mixed populations.

light signals that are proportional to the amount of DNA present in the sample (Figure 5.1). To ensure the amount of human DNA can be specifically measured, the sample is denatured and re-annealed to allow binding to added probe molecules specific for a human repetitive DNA sequence. Non-specific background signals can then be subtracted to give a value for the amount of human DNA.

However, the technique is only suitable for relatively short fragments of DNA, and the relationship between light emitted and DNA concentration is only linear up to $0.4 \, \text{ng} \, \mu\text{l}^{-1}$. In addition, the performance of the assay may be affected by factors affecting the activity of the enzymic reactions, and accuracy is still reliant on the standards used in construction of the calibration curve.

5.6 Primary Methods of DNA Quantification

Primary methods of measurement provide a link in the chain of traceability from the abstract definition of a unit of International System of Units (SI) to its practical use in measurement. Primary methods are defined as those that can be fully described in terms of SI units for both the measurement process and the uncertainty associated with the measurement. In practical terms, such methods yield measurement values directly without the use of any reference standards. It should be noted that primary methods are not intended to replace current methods for routine laboratory use, as they generally require costly equipment and a high level of analyst skill. Rather, in the absence of certified reference materials for quantitative DNA analysis, such methods could be exploited to determine the absolute amount of DNA, and thus to certify standards that could then be used to underpin existing quantitative measurement methods in end-user laboratories.

The International Bureau of Weights and Measures (BIPM)[18] considers several measurement techniques acceptable as primary methods, including gravimetric analysis and isotope dilution with mass spectrometry (IDMS).

DNA Quantification 95

5.6.1 Gravimetric Analysis

In theory a gravimetric approach could be used to determine absolutely the quantity of an amount of DNA. In practice, however, this is not currently achievable because of the large amount of material (> 100 mg), and absolute purity that would be required to permit accurate measurement. However, with the development of highly accurate increasingly miniaturised balances, this approach may become feasible.

5.6.2 Isotope Dilution Mass Spectrometry for Oligonucleotide Quantification

As classical chemical techniques such as gravimetry are poorly suited to DNA quantification because of scale issues, there is a requirement for alternative approaches that retain metrological traceability but are more suited to the analysis of DNA. Recently a method for oligonucleotide quantification has been developed that is based on isotope dilution mass spectrometry (IDMS). In the years since its first use, IDMS has become well established as a route to highly reliable quantitative trace analysis[19] and is now the method of choice for the quantification of analytes in primary standards and high-calibre certified reference materials by many national measurement institutes. A double IDMS approach has been applied successfully to the analysis of oligonucleotides[20] and the principle of the technique is simple, as illustrated schematically in Figure 5.3.

Measurements are made of the sample spiked with an isotopically labelled analogue (sample blend) and of the natural mix of bases spiked with the same isotopically labelled analogue (the calibration blend). An isotope that has low natural abundance such as ^{18}O, ^{13}C, or ^{15}N is ideal for labelling. The use of isotope labels enables the spike and the sample to be analysed and differentiated by virtue of their differing mass alone, as the chemical and physical properties of the analyte are identical. The labelled species should be isotopically pure otherwise this needs to be accounted for in the calculation.

Practically, the sample and isotopically labelled analogue are mixed and it is critical that full equilibrium is achieved prior to analysis. To facilitate analysis, sample separation techniques such as gas or liquid chromatography can be employed prior to mass spectrometry. The quantification software of the mass spectrometer allows the peak areas to be determined, and the ratio of the areas under the peaks is used to calculate the concentration of the sample, since the concentration of the natural spike in the calibration blend is known. For maximal accuracy and low uncertainty, the ratio of label to analyte should be roughly equivalent and at unity. A rough estimate of analyte levels may be required prior to accurate IDMS analysis so that this can be more easily achieved.

In applying IDMS to synthetic DNA quantification, appropriate labelled analogues are required. As labelled deoxynucleotides and deoxynucleosides are commercially available, the approach developed involved complete enzymatic

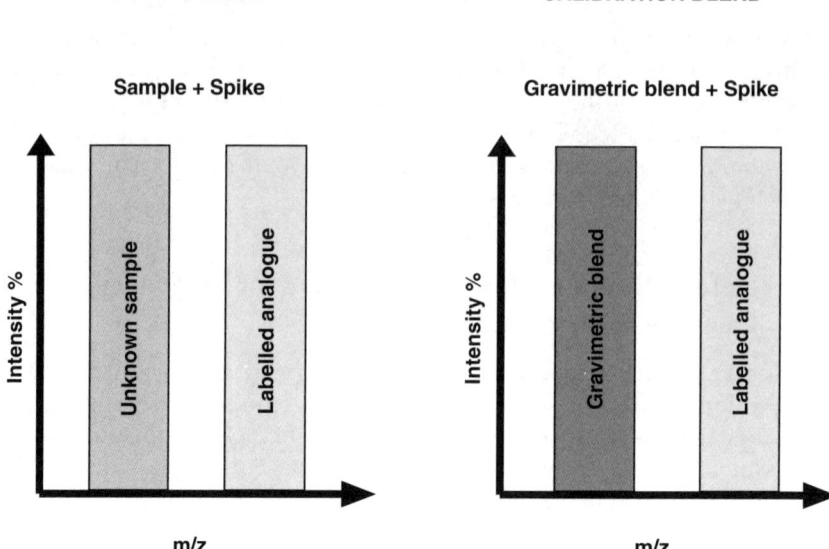

Figure 5.3 Schematic representation of the double IDMS method of DNA quantification. The amounts of each base in the unknown sample are quantified by comparison with the signal from a gravimetrically determined sample. This is enabled by measuring both the gravimetrically characterised and unknown samples simultaneously with a labelled spike blend.

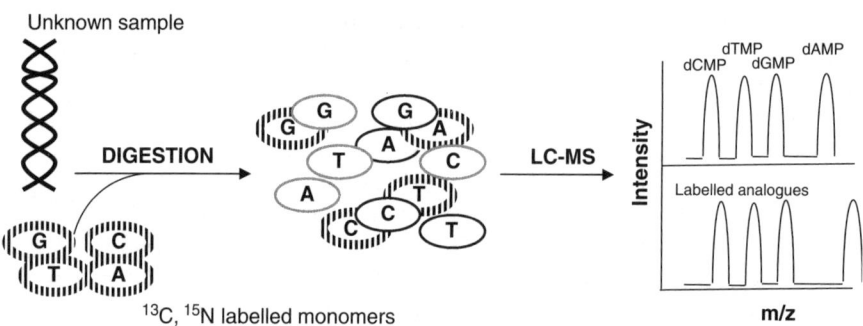

Figure 5.4 Application of IDMS to oligonucleotide quantification.

digestion of the unknown oligonucleotide to its constituent monomer units, and utilising the labelled materials as the internal standard (Figure 5.4).

The accuracy of the method is strongly influenced by the achievement of full equilibrium, effectiveness of the DNA digestion and the gravimetric preparation of the natural analogues. It should be noted that the approach has been developed using relatively small oligonucleotides, and many technical challenges must be overcome before the absolute quantification of larger DNA species such as genomic DNA or PCR products by double IDMS is achievable.

5.7 DNA Quantification by Constituent Phosphorus Determination

An alternative approach, based on phosphorus stoichiometry in nucleic acid, has recently been published[21] and has been shown to give good agreement with the IDMS methodology for oligonucleotides analysis.[22] The principle relies on the existence of a constant amount of phosphorus in the phosphodiester backbone of nucleic acids. In the method the amount of total phosphorus in a DNA sample is accurately determined using inductively coupled plasma optical emission spectroscopy (ICP-OES) and the amount of DNA can then be calculated from this result. The method is represented schematically in Figure 5.5.

Inductively coupled plasma (ICP) efficiently vaporises, excites and ionises atoms by means of very high temperature (7000–8000 °C), and is used as a method of exciting atoms for optical emission spectroscopy (OES). The OES then quantitatively measures the light emission from the atoms, as they decay back to lower energy levels after excitation, to determine the concentration of the analyte. All the atoms in the sample are excited, and so can be detected, simultaneously, which requires a high resolution spectrometer to separate the signals from different elements present. The emission intensity is proportional to analyte concentration at low concentrations ($<10^{-5}$ M), and this property is exploited in quantification. To determine the concentration of an unknown requires calibration of the instrument with a standard, and the quality of the

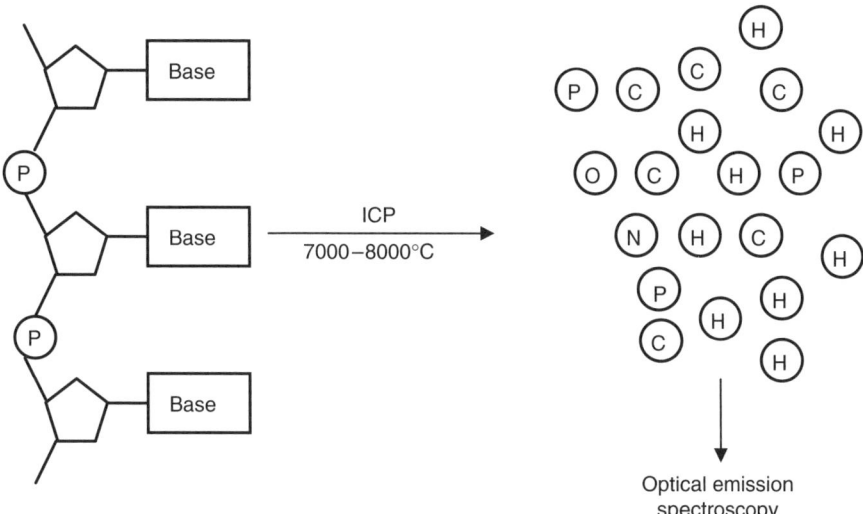

Figure 5.5 Quantification of DNA using ICP-OES analysis. Using a very high temperature plasma support gas, the DNA is nebulised, broken into its constituent atoms and ionised. The phosphorus optical emission spectrum is then measured quantitatively to enable the total amount of phosphorus to be determined by comparison to a known standard.

standard will affect the accuracy of the results. The presence of exogenous phosphorus, incomplete atomisation of the sample and bias in the instrument can also significantly affect the accuracy of quantification, although the error in the method as applied to oligonucleotides has been reported as less than 1%.[21] The method has also successfully been applied to the characterisation of more complex genomic DNA.[2] However, a limitation with this approach is the amount of material required to perform the analysis (~ 1.7 mg per measurement) although this may decrease in the future dependent on technical and methodological advances.

5.8 Comparability of DNA Measurement Methods

The wide range of methods and instruments that may be used for DNA quantification, coupled with the absence of certified quantitative reference materials, means that there may be significant differences in the quantitative results produced by different laboratories, analysts and methods and that these differences are ill-defined.

In-house results comparing the performance of several commonly used techniques, including UV absorbance measurement, fluorescence spectroscopy and ICP-OES analysis, demonstrated that there was poor comparability of results between the methods.[2]

The concentration of a genomic DNA solution was initially determined using ICP-OES analysis, then diluted to allow measurement by two UV absorbance methods (U-2000 and ND-1000) and two fluorescent methods (PicoGreen® and Quant-iT™). The variation of each method from the ICP-OES value and

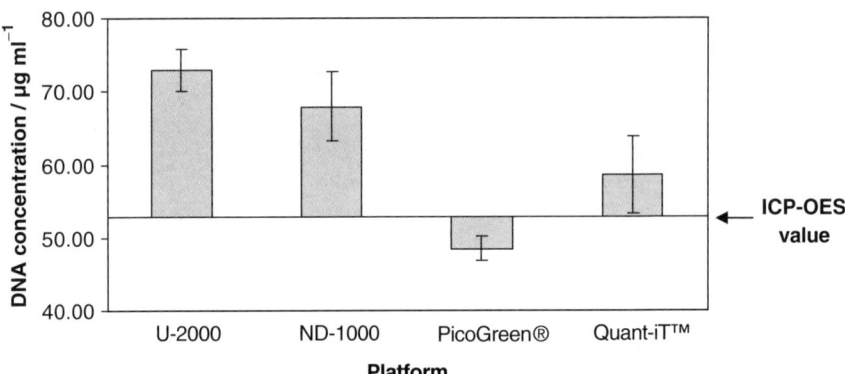

Figure 5.6 Graph illustrating the variability in DNA quantification using both UV absorbance and fluorescence methods, compared to ICP-OES analysis.[2] The quantification methods used were: an in-house ICP-OES technique; U-2000 spectrophotometer (Hitachi, Berkshire, UK); NanoDrop® ND-1000 (NanoDrop, Delaware, USA); PicoGreen® DNA Assays (Cambio, Cambridge, UK); High Sensitivity Quant-iT kit protocol (Molecular Probes™, Invitrogen, Paisley, UK).

the standard deviation of 36 replicate measurements is shown in Figure 5.6. A separate study similarly observed variability between results obtained when quantifying a genomic DNA solution using UV absorbance, a PicoGreen® method and a real-time PCR assay.[1] Variability between RNA quantification methods is discussed in Chapter 9.

5.9 Summary

The requirement for absolute quantification of genomic DNA is fundamental to many molecular analyses, and demands for higher accuracy have increased with the development of quantitative diagnostic assays and legislative requirements in the area of food labelling. Accuracy of determinations are difficult to assess due to the lack of suitable reference materials; reproducibility of determinations, however, can be improved by the employment of good quality samples, trained staff, calibrated equipment, suitable in-house standards and use of methodologies within their limitations (in particular, linear dynamic range).

For routine laboratory use, the most appropriate quantification method will depend on the type of sample being analysed, the volume of material available and the approximate concentration range. Availability of the performance characteristics for currently used methods and instruments is a key tool to enable the most appropriate method to be chosen and, further, to promote appreciation of the level of measurement uncertainty associated with quantitative DNA measurements.

References

1. K. A. Haque, R. M. Pfeiffer, M. B. Beerman, J. P. Struewing, S. J. Channock and A. W. Bergen, *B.M.C. Biotechnol.*, 2003, **28**, 20.
2. C. A. English, S. Merson and J. T. Keer, *Anal. Chem.*, 2006, **78**, 4630.
3. H. J. Aarts, J. P. van Rie and E. J. Kok, *Exp. Rev. Mol. Diagn.*, 2002, **2**, 69.
4. G. Kallansrud and B. Ward, *Anal. Biochem.*, 1996, **236**, 134.
5. A. A. Killeen, *Clin. Lab. Med.*, 1997, **17**, 1.
6. J. Sambrook, E. F. Frisch and T. Maniatis, *Molecular Cloning. A Laboratory Manual*, Cold Spring Harbor Laboratory Press, US, 2nd edn, 1987, ISBN 978-0-879-693091.
7. K. L. Manchester, *Biotechniques*, 1995, **19**, 208.
8. W. W. Wilfinger, K. Mackey and P. Chomczynski, *Biotechniques*, 1997, **22**, 474.
9. G. Lunn and E. B. Sansone, *Anal. Biochem.*, 1987, **162**, 453.
10. F. D. Menozzi, A. Michel, H. Pora and A. O. A. Miller, *Chromatographia*, 1990, **29**, 167.
11. Q. Huang and W. L. Fu, *Clin. Chem. Lab. Med.*, 2005, **43**, 841.
12. V. L. Singer, L. J. Jones, S. T. Yue and R. P. Haugland, *Anal. Biochem.*, 1997, **249**, 228.

13. C. Labarca and K. Paigen, *Anal. Biochem.*, 1980, **102**, 344.
14. A. L. Drobyshev, A. S. Zasedatelev, G. M. Yershov and A. D. Mirzabekov, *Nucleic Acids Res.*, 1999, **27**, 4100.
15. R. K. Saiki, S. Scharf, F. Faloona, K. B. Mullis, G. T. Horn, H. A. Erlich and N. Arnheim, *Science*, 1985, **230**, 1350.
16. R. Higuchi, G. Dollinger, P. S. Walsh and R. Griffith, *Biotechnology (NY)*, 1992, **10**, 413.
17. M. N. Mandrekar, A. M. Erickson, K. Kopp, B. E. Krenke, P. V. Mandrekar, R. Nelson, K. Peterson, J. Shultz, A. Tereba and N. Westphal, *Croat. Med. J.*, 2001, **42**, 336.
18. BIPM, http://www.bipm.org/en/home/ (accessed July 2007).
19. K. G. Heumann, *Fresenius Z. Anal. Chem.*, 1986, **325**, 661.
20. G. O'Connor, C. Dawson, A. Woolford, K. S. Webb and T. Catterick, *Anal. Chem.*, 2002, **74**, 3670.
21. I. Yang, M. S. Han, Y. H. Yim, E. Hwang and S. R. Park, *Anal. Biochem.*, 2004, **335**, 150.
22. C. E. Donald, P. Stokes, G. O'Connor and A. J. Woolford, *J. Chromatogr. B. Analyt. Technol. Biomed. Life Sci.*, 2005, **817**, 173.

CHAPTER 6
PCR: Factors Affecting Reliability and Validity

CHARLOTTE L. BAILEY, LYNDSEY BIRCH AND
DAVID G. McDOWELL

LGC, Queens Road, Teddington, TW11 0LY

6.1 Introduction

The polymerase chain reaction (PCR)[1,2] is an immensely powerful genetic amplification technique offering high levels of sensitivity and specificity in bioanalysis. In the past thirty years there has been a great increase in the number of PCR-based methods available for clinically and environmentally important organisms, for forensic analysis and for research purposes.[3–6] However, in order to achieve optimal performance, great care has to be exercised at all stages of the procedure.

Briefly, the PCR is initiated from two short synthetic strands of DNA, known as primers, which are homologous to opposing ends and strands of a selected double-stranded DNA sequence and delineate the region to be amplified. The specificity of the reaction is dictated by the uniqueness of the priming sequence within the template DNA used in the assay, and control of the stringency under which the primers are allowed to interact with target as opposed to non-target sequences. The sensitivity of the reaction results from sequential rounds of amplification, controlled by a thermal cycler, in which the products of one cycle can potentially act as targets in the next cycle. The process is repeated approximately 30 times allowing the amount of the selected target to be theoretically doubled at each cycle. In practice, the efficiency of the amplification approaches 100% for only a limited part of the process.

Eventually the amount of DNA produced in an amplification reaction will reach a maximum level known as the plateau. Whilst the number of PCR cycles before the plateau that is attained is dependent upon the amount of target at the start of the reaction, the level of plateau (copies of product) is independent of the initial target concentration. Consequently the yield of DNA in reactions amplified to plateau (end-point detection) cannot be used for quantification purposes.

The exponential nature of the amplification process means that subtle differences in amplification efficiency can lead to relatively large differences in product yield and results. Tube-to-tube variation of amplification efficiency can result from pipetting differences between reactions or variations in temperature between different positions within the thermal cycler block. Variations can also occur between different runs on the same thermal cycler, different 'identical' machines and different makes of machine or different batches of reagents. The generation of false positive results due to the presence of contaminating DNA poses an additional threat. There is, therefore, a requirement for both calibration of the thermal cycler and the use of suitable positive and negative controls in order to have confidence in the results obtained.

There are many excellent texts on PCR to which the reader should refer for further detail or more specialised applications.[7–12] Some of the features of basic PCR are given in Table 6.1

Selection of the appropriate parameters for any given application should be based on thorough optimisation of reaction components, the performance of the thermal cycler to be used and the specific question to be answered. Careful consideration and planning of all stages of the amplification process is required if a robust and reliable assay is to be performed.

Table 6.1 Selected features of basic PCR.

Benefits:

- Sensitivity – Can detect very low numbers of target DNA. Theoretically single copy number detection is possible under ideal conditions but substantially higher amounts of target may be required depending upon the type of sample;
- Specificity – Can be used to accurately identify a specific genetic sequence within a complex genetic background such as a whole genome.

Potential Pitfalls:

- Lack of precision due to block-to-block, run-to-run or tube-to-tube variation.
- False negative results may arise as a consequence of inhibition or genetic polymorphism at selected priming sites;
- False positive results due to inadvertent contamination during PCR set-up, possibly due to the generation of aerosols.

Requirements for Effective Use:

- Good planning and experimental design;
- Correct use of suitable controls (positive, negative and non-target);
- Correct use of thermal cycler;
- Consideration of thermal profile and effect on results;
- Good laboratory set-up/housekeeping, equipment calibration;
- Thorough optimisation and validation.

6.1.1 Real-time PCR

Unlike basic end-point detection PCR, real-time PCR utilises fluorescent dyes to enable monitoring of the production of the PCR products in 'real time'. The fluorescence is detected and monitored throughout the exponential phase of the PCR, not the plateau phase of more traditional PCR methods, and therefore generates a signal which increases in direct proportion to the amount of starting template, thus allowing quantification of targets in real time.

Real time PCR will not be covered in the remainder of this chapter, but will be discussed in more detail in Chapter 7.

6.2 The Amplification Protocol

A standard amplification reaction could be 25–100 μl in volume, although volumes as small as 5 μl have been used successfully. The reaction would contain target DNA, deoxynucleotide triphosphates (dNTPs), Mg^{2+}, a suitable reaction buffer, a thermostable DNA polymerase such as *Taq* and oligonucleotide primers defining the start and end points of the final amplified product. A typical 25 μl PCR is illustrated in Table 6.2.

The amplification is normally performed in a small polypropylene tube placed in a thermal cycler, which is programmed to reach typically three selected temperatures in sequence (see Section 6.4). Each cycle of the temperatures is known as a PCR cycle (see Figure 6.1). During each cycle DNA denaturation, primer annealing and primer extension occurs; 25–40 cycles of amplification usually yields sufficient DNA for subsequent analysis by agarose gel electrophoresis, restriction endonuclease digestion, hybridisation or sequencing.

A typical thermal profile for a cycler using a simulated tube or equivalent to assess reaction temperature is given in Table 6.3.

Table 6.2 An illustrative 25 μl PCR reaction set-up checklist.

Reagent	Final concentration	Volume for one reaction (μl)	Volume for 40 reactions (μl) (master mix)	Lot numbers	Check when added
10X Buffer (including 15 mM MgCl$_2$)	1X 1.5 mM MgCl$_2$	2.5	100	#13332	✔
dNTPs (1.25 mM each)	0.2 mM each	4.0	160	#13546	✔
Primer 1 (10 μM)	0.4 μM	1.0	40	#7555-085	✔
Primer 2 (10 μM)	0.4 μM	1.0	40	#7555-086	✔
Taq polymerase (5 U μl^{-1})	1.25 units	0.25	10	#13336	✔
Water	—	11.25	450	Aliquot	✔
DNA	—	5	—	—	✔

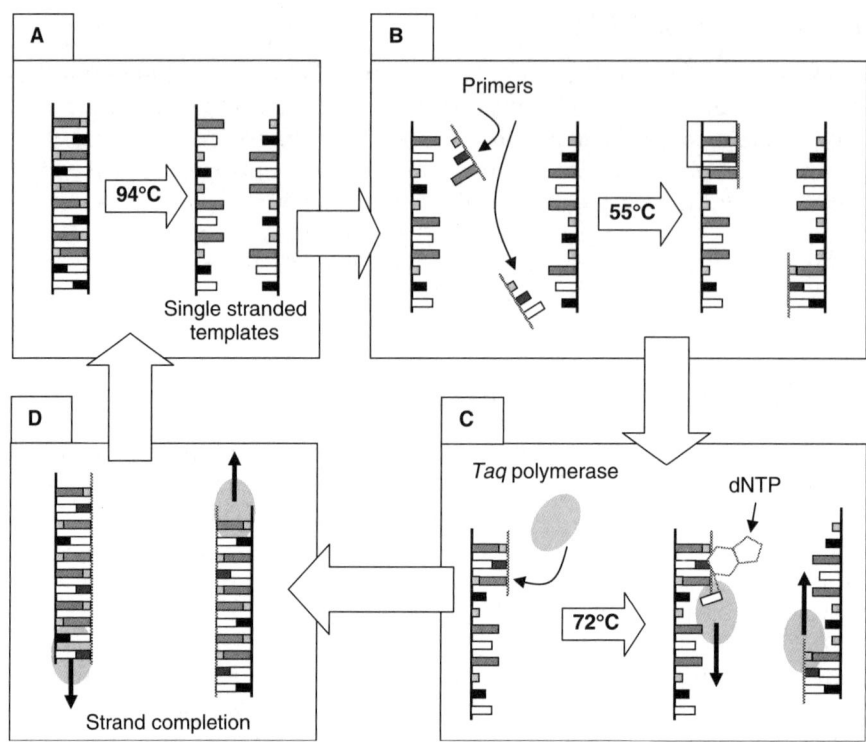

Figure 6.1 Schematic representation of the PCR Process. (A) DNA duplex is separated in denaturation step. (B) Complementary primer sequences are annealed, which flank the region to be amplified – annealing step. (C) Complementary nucleotides are polymerised into the growing strand by the enzyme. (D) Usually extension continues until template sequence has been completely copied.

Table 6.3 A typical thermal profile for PCR amplification.

Temperature	Time	Cycle type	Stage
94 °C	30 seconds	Hold	Initial denaturation
94 °C	30 seconds	30 cycles	Denaturation
55 °C	30 seconds	30 cycles	Annealing
72 °C	30 seconds	30 cycles	Extension
72 °C	5 minutes	Hold	Final extension

Almost every factor associated with PCR can potentially affect the performance of the amplification. In order to generate a robust and specific amplification protocol, a number of factors need to be optimised. Furthermore, if a protocol is to be effectively and repetitively used or transferred successfully between laboratories, great care needs to be taken in the way that such an optimised protocol is employed. The following sections describe some of the parameters affecting performance and reproducibility, which should be considered.

6.3 DNA Template

In the majority of analytical situations, the amount and integrity of the DNA template often limits the type of analysis that is possible. Ideally the DNA template would be of high molecular weight and purified from cellular and additional matrix components, such that it contains no potential inhibitors (this is discussed further in Section 6.4.9). This ideal is, however, often impossible to achieve. Many factors may affect the quality and quantity of the amplifiable DNA template, including the amount of starting material, how representative the initial sample was, the age, storage conditions and processing of the sample and possible presence of inhibitors. All of these factors need to be taken into account when carrying out the PCR. Where limited sample material is available or where samples are aged or processed, PCR is often the only analytical choice, as the sensitivity of the technique means that relatively few targets need to be present or intact in order for analysis to be possible.

6.3.1 Integrity

When amplifying from DNA which may be degraded, it is advisable to select as small a target sequence as is feasible, since these are more likely to have remained intact and will be amplified in preference to larger targets. This is illustrated in Figure 6.2, where the persisting ability to amplify the smaller target (approximately 300 bp) from increasingly degraded template material demonstrates the value of selecting regions of 100–300 bp for successful amplification and analysis of potentially damaged or degraded samples.

6.3.2 Concentration

The concentration of template DNA used in the reaction may significantly affect the results. Too little DNA may give rise to a false negative result, whilst too much DNA may encourage the production of non-specific products seen as artefactual bands on gel analysis, and may even inhibit the reaction.[13] As a guideline, 10^4–10^5 target copies usually results in a good amplification. In principle, lower levels can be used and even single copy detection is possible.

When using template preparations containing very low levels of target DNA, additional precautions may be required such as replicate analyses. Replication could overcome any statistical variation inherent in sampling small aliquots from a larger sample containing low levels of a target. Similarly, sampling variation also needs to be considered when selecting suitable representative starting material for DNA extraction (see Chapter 4). When considering the amount of DNA to be included, it is important to consider the genome size as illustrated in Table 6.4.

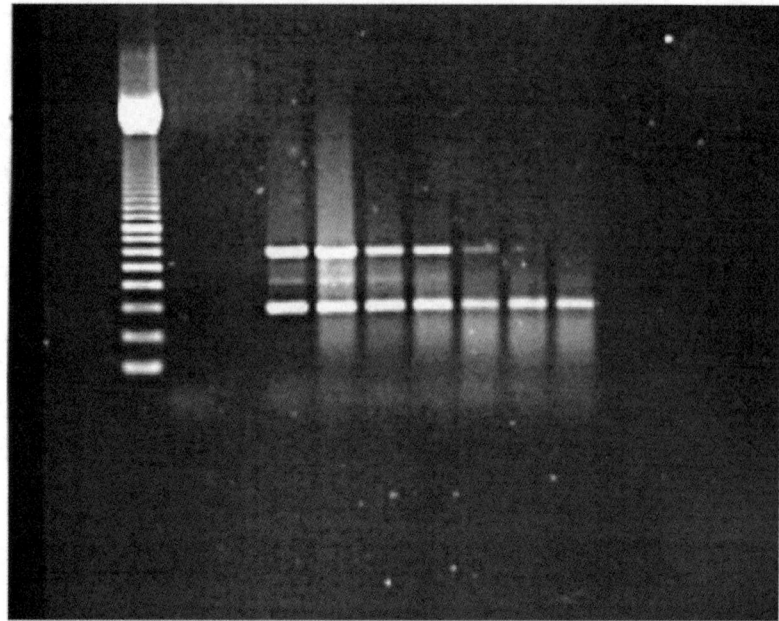

Figure 6.2 The effect of using increasingly degraded DNA as a template for PCR with universal primers. Lane 1, amplification using HMW (High Molecular Weight) calf thymus DNA only, and Lanes 2 to 7, amplification using a mixture of DNA (70% degraded by sonicated + 30% HMW calf thymus). Sonication times were 0 seconds (Lane 2), 10 seconds (Lane 3), 30 seconds (Lane 4), 50 seconds (Lane 5), 70 seconds (Lane 6) and 90 seconds (Lane 7). M represents a 100-bp molecular weight marker and SDW represents a PCR negative control

Table 6.4 Selected genome sizes.

1 μg pGEM® plasmid DNA $=2.85 \times 10^{11}$ copies
1 μg lambda phage DNA $=1.9 \times 10^{8}$ copies
1 μg human genomic DNA $=1.5 \times 10^{5}$ copies (diploid genome)

6.4 Reaction Components and Conditions Affecting Amplification and Reliability

6.4.1 Reaction Buffer

A suitable *Taq* DNA polymerase buffer is usually supplied with the enzyme, to ensure the correct components are present at the appropriate concentrations and at the most favourable pH for optimal enzyme activity. This may comprise a complete buffer (containing MgCl$_2$) or a magnesium free buffer with separate MgCl$_2$ to allow optimisation work to be carried out.

6.4.2 Magnesium Chloride

PCR-based amplification utilises a thermostable DNA polymerase for DNA synthesis. DNA polymerases are magnesium dependent and therefore the major role of magnesium within the amplification mix is to serve as an enzyme cofactor. A number of factors influence the level of magnesium required for optimal reaction performance. A magnesium chloride concentration of 1.5 mM is suitable for many applications and is recommended as a good starting point for optimisation experiments.

When considering the concentration of magnesium, it is important to remember that dNTPs bind magnesium with a stoichiometry of 1:1. In the reaction quoted in Section 6.2, the total dNTP concentration in the final reaction would be 0.8 mM and the level of free magnesium would be 0.7 mM. Optimisation of free magnesium in the range of 0.5 to 2.5 mM is usually sufficient for most applications. As a guideline, too much free magnesium may reduce the specificity of the reaction whilst free magnesium levels of less than 0.5 mM may compromise the activity of the enzyme and reduce the yield of product obtained.

6.4.3 Deoxynucleotide Triphosphates

A deoxynucleotide triphosphate (dNTP) concentration range of 50–200 mM for each dNTP is usually quoted as being optimal for most applications. The concentration can be optimised to maximise specificity although the product yield may be compromised at the lowest levels. The use of high concentrations and unbalanced ratios of dNTPs is likely to compromise the fidelity of incorporation. Again it is important to consider the free magnesium concentration when adjusting the dNTP level.

6.4.4 Water

The quality of the water used in PCR may affect the reaction due to the balance of ions in the water. In general a high quality grade water, such as tissue culture grade or molecular biology grade water, should be used. It is advisable to prepare single-use aliquots of water for use in PCR to prevent contamination.

6.4.5 Primer Design and Target Selection

6.4.5.1 Primer Design

There are several factors which must be considered when designing primers for the PCR assay. Initial considerations focus on the target region for amplification, which may be constrained by sequence information. In highly conserved regions there may be very specific areas of uniqueness which must be exploited for primer binding to ensure the specificity of the assay. Or alternatively in a degenerate assay, the conserved sequences may limit the regions in which PCR primers may be targeted. An alternative is to design primers which may possess one or more different bases at a particular position of the sequence, or to incorporate inosine at particular positions within the primer to allow binding to any nucleotide.

If amplification of a single target is required, primers should be checked against sequence databases for uniqueness. Checks to ensure that the primer does not self-hybridise during the annealing stage, or bind to the other primer, are essential, as both these interactions can significantly affect the overall efficiency of amplification.

A further aim is to roughly match the melting temperature (T_m, the temperature at which 50% of the template annealing sites and primers are in duplex) of the primers used in an assay, to ensure that both primers bind to the template with similar efficiency. If the T_m of one of the primers is much higher than that of the other, then an asymmetric reaction may occur resulting from the preferential binding of one of the primer pair. However, if a single-tube nested PCR (a PCR in which the initial product is re-amplified using a second primer set) is required then the two primer pairs may be designed with quite different melting temperatures to allow initial production of a longer amplicon with a higher annealing temperature for several cycles, followed by a lower annealing temperature in later cycles to produce the shorter, nested product. Guidelines which may help to achieve success in primer design are given in Table 6.5.

A number of computer packages are also available, for example the Primer-Select module of Lasergene[14] and Primer Express®[15], which will assist in primer design but do not guarantee success. Given the low cost of primer synthesis, in practice it is often advisable to synthesise and test more than one primer pair as the actual performance of an assay cannot always be predicted.

Whilst the major aim of primer design is to select the target, the primers can often be modified for additional purposes. For example, degenerate PCR can be used when only a protein motif is available, and there is a need to back-translate the protein motif to the corresponding nucleotide sequence. Instead of using specific PCR primers with a single sequence, mixed PCR primers are used and 'wobbles' are inserted into the PCR primers where there is more than one translational possibility (see Figure 6.3).

Table 6.5 Guidelines for primer design.

- Select a primer size of 17–30 bases.
- Balance the predicted T_ms of the primer pair (best range 55–80 °C).
- Select GC content of primers of between 45 and 60% (if possible).
- Avoid complementarity of the primer pair, particularly at the 3' end.
- Include C or CG at 3' end of primers to increase efficiency, however more than 3 CG's at 3' end may promote mis-priming in CG-rich regions.
- Avoid runs of purines, pyrimidines or repetitive sequences
- Avoid regions possessing significant secondary structure.

                    Asp   Ala   Gln   Trp   Gly   Thr

            5'   GAY   GCN   CAR   TGG   GGN   CAN-3'

Figure 6.3 Example of a degenerate PCR primer designed to a protein motif. In the diagram, the Y=C or T, R=G or A, N=G, A, T or C.

For cloning purposes extra sequences may be added at the 5' end of the primer to incorporate restriction endonuclease cleavage sites,[16] or to serve as further priming sites of use in the construction of mimics in competitive PCR.[17] The design of primers with base changes to the original sequence can be used for site directed mutagenesis[18] or the introduction of promoter sequences to allow *in vitro* transcription.[19,20] A range of chemical modifications is also available for standard oligonucleotide synthesis, which may be exploited in PCR for capture and detection purposes (for example addition of biotin, fluorophores and digoxigenin).

6.4.5.2 Target Selection

When selecting a primer pair, it is also important to consider the size of product to be amplified, although a target of approximately 100–200 bases is generally considered to offer a reasonable choice of sequence for the design of primers.

Very small targets (below 100 bp) will allow the use of shorter extension times and increase the speed of analysis. However, it is also important to be able to distinguish between the desired product and any primer dimer formations when analysing results using standard electrophoresis.

Larger target sizes may be required if the product is to encompass suitable restriction sites for subsequent analysis or for sequence investigations. For particular purposes, it may be necessary to amplify substantial lengths of DNA.

Multiplex PCR may require a range of sizes to be targeted if identification is based on the use of electrophoretic separation. This is discussed in more detail in Chapter 8.

6.4.6 Thermostable DNA Polymerases

The polymerase chain reaction was first demonstrated using the thermolabile Klenow fragment of *E. coli* DNA polymerase.[1] The original process was very laborious since the DNA polymerase enzyme was not heat stable and consequently required the addition of fresh enzyme after each denaturation step. It was not until amplification was demonstrated using a thermostable DNA polymerase from *Thermus aquaticus*[2] (*Taq* polymerase) that the true potential of the modern PCR process was realised. The initial use of the polymerase from *Thermus aquaticus*, *Taq*, allowed the cyclical reaction of the PCR to proceed without the addition of enzyme at every cycle.

Since the initial use of *Taq* DNA polymerase, many other DNA polymerases have been discovered and tested for use in PCR and there is now a number of thermostable DNA polymerases which can be used (see Table 6.6). There are many different types on the market from native polymerases and recombinant polymerases to chemically modified ones, and also enzyme blends for particular applications. These modifications convey even more advantages to the user than the unmodified enzymes, such as hot-start capabilities (see Section 6.4.7).

Taq polymerase is still one of the most common choices and is usually used at 0.05 U μl^{-1}. This is an excess of enzyme for most purposes. The use of additional enzyme to improve product yield is unlikely to be successful and will increase the probability of generating non-specific products seen as artefactual bands and smearing on gel analysis. An excess of *Taq* can also inhibit the reaction.[13]

6.4.6.1 Factors Affecting Choice of Polymerase

The factors affecting the choice of enzyme are summarised below:

- What degree of thermal stability is required?
- What processivity is acceptable?
- Is a proof reading capability required (3'-5' exonuclease)? For example for cloning and sequencing, long PCR or where internal regions may cause misincorporation and premature termination?
- Is a lack of proof reading capability required? A primer with a 3' mismatch would not normally be expected to extend but would with such a 3'-5' exonuclease activity;
- What range of optimal magnesium concentration is required?
- Would the lack of a 5'-3' exonuclease activity be beneficial? For example, where more product/higher plateau is required such as in multiplex PCR or RAPD analysis;
- Is an associated reverse transcriptase activity required, for example in RT-PCR?

6.4.7 Hot-start Mechanisms

Hot-start mechanisms are used to increase the specificity of the PCR by preventing reactions occurring before a specified temperature is reached thus reducing mis-priming events. There are several commercially available hot-start approaches such as AmpliTaq Gold®[15] and SureStart™ Taq DNA Polymerase.[22,23] Reported approaches utilise a number of methods to prevent incorrect extension prior to the first amplification cycle, such as wax (see Section 6.4.7.1) and antibody (see Section 6.4.7.2) to prevent premature amplification. All hot-start PCR mechanisms should be effective in reducing non-specific amplification.

Table 6.6 Characteristics of commercially available DNA polymerases.

DNA Polymerase	Activity
Taq Polymerase	Thermostable – half life 45–96 minutes at 95 °C (9 min at 97.5 °C) Extension rate is: <60 n/s @ 70 °C 25 n/s @ 55 °C 1.5 n/s @ 37 °C Processivity – moderate (50–60 nucleotide extension before dissociation) Fidelity – Between 3×10^{-4} and 3×10^{-6} errors per polymerised nucleotide 5'-3' Proof reading exonuclease activity No 3'-5' exonuclease activity (errors can be introduced)
Stoffel Fragment (a deletion derivative of *Taq* Polymerase)	Thermostable – half life 80 minutes at 95 °C (21 min at 97.5 °C) Extension rate – approx. 50 nucleotide per second at 70 °C Processivity – Low (5–10 nucleotide extension before dissociation) No 5'-3' proof reading exonuclease activity No 3'-5' exonuclease activity Works under broader range of magnesium concentrations than *Taq* Useful when optimising multiplex reactions
Pfu Polymerase	Thermostable – One of the most thermostable enzymes Extension rate – Lower extension rate than *Taq* polymerase Processivity – Low 5'-3' Proof reading exonuclease activity 3'-5' Exonuclease activity Lowest rate of error incorporation 11–12 Fold greater replication fidelity than *Taq*[21]
VENT Polymerase	Thermostable – half life 7 hours at 95 °C (greatest thermal stability) Fidelity – 5–15 fold higher than that of *Taq* Polymerase 3'-5' Proof reading exonuclease activity
Topo Taq (high fidelity system)	Hybrid DNA polymerase It is linked with unique non-specific DNA binding domains which increase processivity, thermostability and specificity High yield amplification and more robust Ability to amplify 20 kb template in the absence of exonuclease activity Enhanced by the addition of a hyperstable Methanopyrus DNA topoisomerase that facilitates DNA strand separation 5'-3' Proof reading exonuclease activity 3'-5' Exonuclease activity

6.4.7.1 Wax

There are many different approaches on the market today utilising wax as a means of hot start. They all use layers of wax to physically separate the polymerase, template or specific reaction components (for example magnesium) from the remaining reaction components. Other methods contain one reaction component inside wax beads, for example AmpliWax® PCR Gem[15] and Promega Taq Bead™ HotStart Polymerase.[24] In both the wax layer and beads systems all the reaction components mix only when the wax melts during the first denaturation step and this melting occurs at a high enough temperature to prevent mis-priming events, thus reducing the amount of primer dimer and non-specific amplification products generated.

6.4.7.2 Antibody

A *Taq* specific monoclonal antibody can be added to the assay to inactivate the DNA polymerase. This will then become heat inactivated itself during the first denaturation cycle, allowing the reaction to proceed. There are currently several products on the market including: TaqStart™ Antibody,[25] Jumpstart™ Taq Antibody,[26] TaqStart™ Antibody[26] and Platinum® Taq Antibody.[27]

6.4.8 PCR Optimisation

There are many factors that can affect the sensitivity and specificity of a PCR. When developing a new assay, it is important to optimise the reaction to achieve the desired result. Traditionally, the effect of changing each parameter on the PCR would have to be assessed as outlined in Table 6.7; this is a laborious task. However, many papers, manuals and manufacturers give detailed guidelines for PCR optimisation procedures. In addition, methods are available for varying a number of parameters simultaneously and use mathematics to calculate the best combination of factors from the composite data.[28] These methods, which expand a modified Taguchi method for PCR optimisation,[29,30] are less labour intensive.

When optimising a PCR, magnesium concentration is usually the first variable to be tested, as changing the concentration can affect both reaction specificity and sensitivity. The level of dNTPs and the annealing temperature can also affect the efficiency and specificity of the PCR, therefore both require optimisation. In terms of the annealing temperature, it is advisable to start at the low end of the T_m range to be tested, which is determined by the primer T_ms, and increase the temperature stepwise in gradual increments as necessary. If the temperature is too low, the reaction will be non-specific. However, if the temperature is too high, the stringency may affect reaction efficiency, resulting in no amplicon production, or very poor yields. It should be noted that other reaction components, including template DNA, chelating agents such as EDTA, and proteins can also affect the amount of free magnesium. Different enzymes may also require different buffer conditions and it is therefore

Table 6.7 Factors to consider when optimising PCR performance in terms of sensitivity, specificity and fidelity.

Parameter	Adjustment	Effect on Sensitivity	Specificity	Fidelity
Mg^{2+}	↑ ↓	↑ ↓	↓ ↑	
dNTP concentration	↓ ↑			↑ ↓
Annealing temperature	↑ ↓	↓ ↑	↑ ↓	
Cycle number	↑ ↓	↑ ↓	↓ ↑	
Hot start	+ −	↑ ↓	↑ ↓	
Proof reading enzyme	+ −			↑ ↓

important to check the manufacturer's guidelines before using a polymerase for the first time.

6.4.9 Inhibitors and Enhancers

The polymerase chain reaction[2] can be inhibited or enhanced by many different substances (outlined in Tables 6.8 and 6.9), arising from the raw biological sample or the method and reagents used to extract the DNA.

A wide variety of biological samples are used for PCR, including animal tissues and bodily fluids, bacterial samples, forensic and archaeological material and plant tissues. Moreover, many of these may be sourced from crude environmental samples, for example foodstuffs, soil and sludge. Many of these crude preparations contain substances inhibitory to PCR,[31] although the exact mechanisms of inhibition and whether there might be an antagonistic effect between individual components is not always known.

Assuming that only a fraction of any given sample or extraction solution is transferred to a PCR, only a few substances will reach concentrations where they are inhibitory on their own. Individual compounds may, however, become concentrated by the extraction method, for example by co-precipitation with the sample DNA. The inhibitory mode of action of some individual compounds might be linked with precipitation of the DNA, denaturation of DNA or the polymerase enzyme, binding of the necessary Mg^{2+} ions or adding an excess of Mg^{2+}. Many substances may have an assay-specific effect; at certain concentrations there is inhibition/enhancement, but transfer these conditions to a

Table 6.8 Suggested mode of action of some common PCR inhibitors.

Reagent/Sample	Inhibition details	Comments
Molecular Biology Reagents		
Ionic detergents	Very inhibitory	Effects of low concentrations can be reversed/neutralised by certain non-ionic detergents (e.g. 0.5% Tween 20 and Nonidet P40)
SDS (sodium dodecyl sulfate)	Range > 0.005, <0.01% (v/v) (none at 0.005%), reducing *Taq* polymerase activity to ≥ 10%	
Non-ionic detergents (Triton X-100, Tween 20, Nonidet P40)	Non inhibitory up to 2% (v/v)	Tween 20 10% (none at 2%) (v/v)
		Triton X-100 2% (none at 1%) (v/v) Nonidet P40 > 2% (none at 0.2%) (v/v)
Sodium hydroxide	Inhibitory > 8 mM	Presume due to pH-mediated denaturation of enzyme and dsDNA
Alcohols	Ethanol > 5% (none at 2.5% v/v) Isopropanol > 1% (v/v) (none at 0.5% v/v)	Possible DNA precipitation
EDTA (Ethylenediaminetetraacetic acid) and EGTA (ethyleneglycol-bis (βaminoethyl ether) tetraacetic acid)	Inhibitory > 1 mM Can also be used as an enhancer at low <1 mM as it stabilises the polymerase	EDTA known to chelate Mg^{2+} ions
Proteinase K	Inhibitory	Digestion of *Taq* DNA polymerase
Urea	Inhibitory at 0.5 M	Effectively removed by simple dialysis or ultrafiltration

Table 6.8 (*Continued*)

Reagent/Sample	Inhibition details	Comments
Formamide	Inhibitory $\geq 5\%$ (v/v)	Possible reduction in the activity of *Taq* polymerase
Tetramethylammonium chloride (TMAC)	>0.5 mM (some inhibition)	
$MgCl_2$	Inhibitory >0.5 mM	
LiDS (Lithium dodecyl sulfate)	Inhibitory $>0.01\%$ (v/v) (none at 0.005% v/v)	
BSA (Bovine Serum Albumin)	Inhibitory 25 mg ml^{-1} (none at 10 mg ml^{-1})	Can act as an enhancer when used at 10 μg ml^{-1}–100 μg ml^{-1}, as it may bind PCR inhibitors carried through from extraction
Spermidine	Inhibitory 1 mM (none at 0.1 mM)	
Ethidium Bromide	Inhibitory 1% (v/v) (none at 0.1% v/v)	
Acetonitrile	Inhibitory 10% (v/v) (none at 2.5%)(v/v)	
Guanidinium isothiocyanate (GITC)	Inhibitory 100 mM (none at 20 mM)	
Cetyltrimethylammonium bromide (CTAB)	0.1% (v/v) (none at 0.001% v/v)	
Gel dyes, for example bromophenol blue and xylene cyanol	Most inhibit, although some have no effect	Cresol red and tartrazine are non-inhibitory
UV damaged mineral oil	Inhibit	Oil treated with UV overnight inhibits due to free radicals produced
Clinical samples		
Heparin	Inhibitory. Naturally occurs in liver, lung and artery walls, and used as anticoagulant for blood samples	Heparinase can be used to eliminate the heparin

(*Continued*)

Table 6.8 (*Continued*)

Reagent/Sample	Inhibition details	Comments
Other blood components	Inhibitory. Possibly porphyrin compounds	Overcome by lysing red blood cells and centrifuging DNA containing white cells
Haeme compounds	Inhibitory	Solved by addition of BSA, presumably binding the haeme compound
Cervical specimens	Inhibition	Partly correlated to pH of specimen
Intraocular fluids (viral detection)	Inhibition from ≥ 20 μl aqueous and ≥ 0.5 μl vitreous intraocular fluids in a 100 μl PCR reaction[34]	Overcome by phenol:chloroform extraction
Faeces	Inhibitory, probably through complex plant polysaccharides	Overcome by extraction in presence of cationic surfactants, for example Catrimox-14
Food samples		
Milk, cheese	Possible presence of inhibitory components, including proteases	Enzyme degradation[35,36]
Dairy	Inhibition from Ca^{2+} ions	Competition between Mg^{2+} and Ca^{2+} ions for the polymerase binding site[37]
Microbial samples		
Culture media	Varying inhibitory effects depending on medium used	
Wooden toothpicks	Inhibition in reactions with low Taq levels, when used to transfer bacterial colonies directly to PCR reactions	May be overcome by the use of plastic colony picks, or pipette tips

Table 6.8 (*Continued*)

Reagent/Sample	Inhibition details	Comments
Fungi	Inhibitory polyphosphates	No real data or explanation, thought that polyphosphates may be responsible for difficulties.
Environmental samples		
Soil	Humic acid, fulvic acids, heavy metal ions. Possible mode of action through binding of polymerase or target DNA	Inhibition overcome by adding any of the following to reaction: carbonic anhydrase, ovalbumin, BSA or myosin. BSA is most suitable. Alternatively inhibitors may be removed by sephadex column purification, and humic acid inhibition relieved by calcium chloride precipitation
Pollen	Inhibition; pollen contains many biological substances, including enzymes	Enzymatic digestion possible cause

different PCR and there may be no effect or the opposite effect.[32] In addition, the effect of a substance may vary depending on the polymerase used in the assay,[33] and thus inhibition may be overcome simply by changing the enzyme.

Tables 6.8 and 6.9 illustrate some common examples of inhibitors and enhancers of the PCR process and outline their modes of action.

It is possible to try a combination of measures to improve PCR performance. An alternative is to use the commercially available Failsafe kit,[38] which provides consistent high-fidelity PCR results for a wide range of DNA templates and assays. It includes an enzyme mix and 12 PCR pre-mixes that cover a matrix of enhancer and reaction conditions that may be used to find the optimal conditions for a specific PCR. It is claimed that the kit is able to overcome most PCR-related problems. PCR inhibition may be caused by a variety of substances, highlighting the need for more routine inclusion of appropriate internal reaction controls[39] to identify false negative results caused by PCR inhibition.

Table 6.9 Suggested mode of action of commonly used enhancers of PCR.

Action	Enhancement
Use of matrix specific extraction protocol	Specifically designed to increase DNA yield from specific matrix and prevent known inhibitors being co-purified
Perform additional DNA clean-up	Removes inhibitors by employing additional purification procedures
Dilution of DNA-containing solution	Reduces concentration of inhibitors added to a PCR reaction. Only appropriate if a sufficient concentration of DNA is available
Heat treatment (5–15 minutes at 95–100 °C)	Inactivates proteases and DNases present in DNA extract
Increase *Taq* DNA polymerase concentration	May overcome inhibition due to enzyme inactivation or by successfully competing with agents that chelate essential enzyme co-factors
Use alternative DNA polymerase enzyme	Alternative DNA polymerases may be affected differently by inhibitors or may allow the use of higher denaturation temperatures
Add DNA-stabilising cosolvent such as TMAC	Increases T_m by stabilising AT base pairs. At 15–100 mM, may be useful for reducing non-specific products from AT-rich targets by allowing the annealing temperature to be increased[34]
Add DNA-destabilising cosolvent, such as DMSO, formamide, betaine or glycerol	Improves the amplification of GC-rich targets by reducing the T_m of DNA. Betaine (1.0–1.7 M) stabilises AT base pairs whilst destabilising GC interactions. DMSO (2–10% v/v) reduces secondary structure). Glycerol (5–10% v/v) reduces T_m
Add non-ionic detergents, such as Tween 20 or Nonidet P40	At 0.15–1% (v/v) can reduce secondary structures and stabilise *Taq* DNA polymerase
Add protein agents (BSA, gp32)	Can quench protease activity or preferentially bind inhibitors
Decrease MgCl$_2$ concentration	MgCl$_2$ strongly stabilises the DNA duplex therefore a reduction may help achieve complete denaturation

Table 6.9 (*Continued*)

Action	Enhancement
Increase $MgCl_2$ concentration	Can increase the amount of free magnesium, and preferentially compete with inhibitory ions for *Taq* DNA polymerase binding site
Add polyamines (spermine)	Stabilises DNA and possibly stabilises enzyme activity
Substitute nucleotide with analogue (c7dGTP)	Reduces secondary structure and non-specific product formation

6.5 Thermal Cycling

6.5.1 Cycle set-up

Although the process of PCR is described as three discrete stages, a more realistic view of the amplification reaction is as a continuous cycle of events (Figure 6.1) encompassing denaturation, annealing and extension as described below.

6.5.1.1 Denaturation

A denaturation temperature between 94 and 96 °C is commonly used to denature the template DNA, however high temperatures will gradually denature the polymerase. In later stages of the PCR the denaturation temperature may be reduced, as the size of the majority of the duplex target is smaller. Alternatively, higher temperatures for shorter lengths of time may also be effective in increasing the number of cycles that can be performed before appreciable loss of enzyme activity has occurred.

6.5.1.2 Annealing

The annealing temperature is chosen dependent on the calculated T_m of the primers, and is generally about 5–10 °C lower than the lowest T_m of the primer pair. A variety of methods can be used to calculate T_m theoretically, and thus it is important to ensure that the T_ms of primer pairs are calculated in the same way. Too low an annealing temperature may cause lowered specificity of the reaction, while too high an annealing temperature will reduce the efficiency (and thus the yield) of the PCR. Unless the primers are very long, an annealing time of 0.5 minutes is sufficient.

6.5.1.3 Extension

Taq polymerase from *Thermus aquaticus* works optimally at 72 °C, although it does have a reduced activity at 37 °C and below. As a rule of thumb, 1 minute of extension is allowed for formation of each kb of amplicon, with most reactions using extension times of between 0.5 and 3 minutes. Longer times may be required to amplify longer targets, and may also be helpful in later cycles when reaction components (dNTPs and/or primers) may become limiting.

6.5.1.4 Cycle Number

The more amplification cycles that are performed, the higher the yield of the reaction, as long as the components of the reaction are not completely limiting and the enzyme retains some activity. Through the repeated denaturation cycles of the PCR the enzyme activity does become depleted, and in practice running more than 40 cycles does not appreciably increase yield. It is also important to balance speed/throughput with overall efficiency, so it may be effective to use a smaller number of amplification cycles, if this allows an additional run to be performed within the working day.

6.5.2 Thermal Cycler

The reliability of a PCR can depend heavily on the performance of the thermal cycler used. PCR is a stochastic process, and small changes in reaction conditions during the early cycles can greatly affect the overall efficiency of the reaction.

Correct thermal profiles are central to obtaining reliable results, and a number of guidelines can help in achieving this reliability:

- Regularly check the temperature reached by the block – usually achieved through calibration by manufacturer;
- Check uniformity of block temperature – also through manufacturer or other external calibration;
- Keep block clean, in accordance with manufacturer's guidelines;
- Consider the effect of ambient temperature – check manufacturer's guidelines;
- Ensure any sample tubes with temperature probes are comparable to reaction tubes (different solution volumes, or differences in tube wall thickness, will affect the profile if using a sample-controlled thermal cycler);
- Ensure the thermal cycler is used in accordance with the manufacturer's instructions.

The speed with which the thermal cycler changes between the cycle temperatures is known as the ramp rate, and this can be controlled on many instruments.

6.5.3 Temperature Control

The manner in which the thermal cycler is set up to achieve the correct temperature varies and this also needs to be considered if the machine is to

PCR: Factors Affecting Reliability and Validity   121

be used correctly and in order to ensure thermal profiles applied to different machines are comparable.

A positive control kit, the 'OK Kit' from Microzone,[40] is designed to produce three products of varying intensities, depending on the thermal cycling conditions. The kit may be used to ensure that the thermal cycler has reached the correct conditions during the cycling programme. Acceptable results from the kit demonstrate that the thermal cycler has performed as expected, and therefore any anomalies in the experiment are due to another variable.

6.5.3.1 Block Control

In machines using block control, the time that the block is maintained at a given temperature is measured. In such instances, hold times need to be relatively long in order to allow the sample to equilibrate with the block temperature. With block control the use of short annealing times may improve specificity. Improvement may result either from a reduction in the time available for mis-priming events to occur, or from a failure of the reaction to attain the set annealing temperature in the time available (which consequently and inadvertently results in the use of a higher and more specific annealing temperature). Inversely the use of short denaturation times should be avoided as this may result in failure to achieve complete denaturation.

6.5.3.2 Reaction Control

Shorter hold times can be used where the hold time measured is the time the reaction (rather than the block) is predicted to be within a certain temperature window. Whilst such a window is often stated to be within 1 °C of the set temperature, this is not always the case and is a further hidden variable between different thermal cyclers. As the reaction temperature approaches the set temperature, two routes to achieving the final temperature may be taken.

1. The rate of temperature change decreases allowing the sample temperature to catch up. The rate of temperature change and the temperature from which the new rate of change comes into effect will determine the time the sample remains within the given temperature hold window. This may result in differences in effective annealing, extension and denaturation times for different thermal cyclers, even when the same settings are used;
2. The block overshoots the set temperature for a given time before dropping back to the set temperature. The effect of this is to give a more rapid transition to the set temperature which may again result in differences between machines.

6.5.4 Ramp Rate

It is also possible, in many instances, to specify the speed of progression from one temperature to the next, otherwise known as the temperature ramp rate.

This is likely to have little effect for many amplification applications, but the exceptions may include short targets with short extension times, where the extension occurs mainly during the ramp. Thermal cyclers having ambient or sub-ambient temperature capabilities will ramp at different rates irrespective of the set rate, since ambient machines will be affected by room temperature which varies from day to night and from season to season. Sub-ambient machines cooled by pumped water will be similarly affected by water temperature and flow rate. Depending on the way the machine is designed to function, there may be either a longer than predicted cycle time, or failure to attain the set temperature before progressing to the next stage of the cycle. This problem has largely been overcome by the introduction of Peltier heating blocks in newer thermal cyclers.

6.5.5 Alternative Thermal Cyclers

Over the last few years, manufacturers have sought to minimise amplification times required in PCR. Such thermal cycling systems use small reaction volumes in glass capillaries (LightCycler®[41]), giving large surface-area-to-volume ratios, which result in almost instantaneous temperature equilibration and minimal annealing and denaturation times. This, accompanied by ramp rates of 10–20 °C s^{-1} made possible by the use of turbulent forced hot air systems to heat the sample, results in an amplification reaction completed in tens of minutes.

Such systems are often coupled with real-time capabilities, for example the iCycler® thermal cycler[42] and the Rotor-Gene™ 6000,[43] using fluorescent detection of amplification products as the reaction progresses. Again with such technologies it is important to correctly optimise the system since the amount of time spent within a given temperature window is minimal if used at full speed, and there is little chance of annealing occurring during the ramp to and from an incorrectly selected temperature.

6.6 Contamination Control

The polymerase chain reaction is a very powerful genetic amplification tool. Theoretically able to detect a single DNA target by producing approximately 10^{12} copies of a selected sequence in only a few hours, the technique may easily be a victim of its own success. All the products of a PCR amplification are prime candidates for re-amplification and could potentially give rise to false positive results if not excluded from subsequent amplifications.

The risk of potential carry-over contamination should not be underestimated. By illustration, 10^{12} molecules of amplified product diluted in an Olympic-sized swimming pool containing 2500 m^3 of water would result in four amplifiable molecules μl^{-1}.[44]

Potentially as serious, although perhaps more surprising, is the threat posed by unnecessarily concentrated positive control DNA stocks. For instance, a 2.5 kb plasmid control with a concentration of 1 μg μl^{-1} contains 3.6×10^{10} copies

Table 6.10 Precautions for minimising the potential for PCR contamination.

- Use dedicated areas and equipment, including a laminar flow hood
- Use filter-guarded pipette tips or positive displacement pipettes
- Use DNA controls at sensible concentrations
- Use gloves, and change them regularly, to prevent cross contamination of samples
- Prepare large stocks of reagents and freeze in single-use aliquots
- Include negative controls for both PCR set-up and the whole analytical process (this will identify the area of contamination)
- Use highly purified DNA polymerase to avoid false positive amplification from bacterial contaminants
- Consider the use of UNG and dUTP to allow reactions to be purged of previously amplified product
- Consider the use of appropriate decontamination procedures for equipment and benches (UV, sodium hypochlorite)

per $0.1\ \mu l^{-1}$. Aerosol formation from concentrated DNA solutions may represent a severe contamination risk to other reactions. The centrifugation of DNA-containing tubes prior to opening can help to avoid splashing/aerosol formation, whilst keeping tubes closed and the frequent changing of gloves may reduce the risk of contamination.

Sample-to-sample contamination prior to DNA extraction may not necessarily represent as great a risk since a fairly gross contamination and/or extensive amplification would be required to give a false positive result. However, such a theoretical consideration should not be used to endorse casual working practices as, in many cases, absolute confidence in the results is paramount.

A different category of contamination may originate from the polymerase used for the amplification. The enzyme is typically purified from a bacterial source and can contain contaminating DNA at levels estimated at 100 copies per unit of enzyme.[45] This may be a problem where there is homology between the primers selected and the contaminating DNA. This is particularly the case when amplifying a multicopy sequence such as the 16S rRNA genes using primers to highly conserved regions prior to sub-classification. Precautions for minimising the potential for contamination of PCR amplifications are given in Table 6.10.

A number of working practices and procedures exist for excluding unwanted DNA targets from amplification reactions and are discussed below. It is important that these are considered before work starts since it can be difficult or expensive to resolve contamination problems after the event.

6.6.1 Physical Laboratory Separation and Dedicated Equipment

The laboratory area needs to be controlled in a way that matches the analytical requirement. It is important to separate any pre-PCR areas and the PCR reaction set-up from the PCR positive analysis areas of the laboratory. If sufficient space is available it is desirable to maintain dedicated areas and equipment for pre-PCR sample manipulation and storage, PCR reaction set-up, template addition and post- PCR analysis. The key aim is to ensure that PCR set-up and all associated reagents and consumables can be maintained in a DNA-free environment.

This physical separation means that certain rules are in operation within the laboratory in order to ensure that contamination is kept to a minimum:

- Each area has its own dedicated set of equipment, which should not be removed from the area and may be colour coded for ease of identification;
- Some of the areas have specific laboratory coats to be worn, and again may follow the colour-coding scheme. Laboratory coats, worn in the PCR-positive parts of the laboratory, should not enter the pre-PCR areas. The dedicated laboratory coats should not leave each designated area;
- Laboratory notebooks and pens, for instance, should not be taken into the clean PCR set-up area if they have previously entered the PCR-positive areas of the laboratory, as they may carry contaminating PCR products or template DNA.

This is discussed further in Chapter 2.

6.6.2 Pipettes

The use of positive displacement pipettes or filter-guarded pipette tips when handling any DNA solution will prevent contamination of pipette barrels as a result of aerosol formation.

6.6.3 Methods of Decontamination

A range of methods exist which seek to control contamination by destroying unwanted targets. Since prevention is better than cure, they are not recommended as a substitute for good working practice, but may be of benefit where additional levels of confidence are required, or where local working practices may of necessity compromise other control measures.

6.6.3.1 *Uracil-N-glycosylase and dUTP*

This method is specific for degrading amplified products containing dUTP. Substituting dTTP with dUTP in amplification reactions makes it possible to specifically degrade previous amplification products by the use of uracil-N-glycosylase (UNG).[46] In practice, the PCR is set up with the inclusion of UNG. A preliminary incubation at 50 °C allows the UNG to destroy any previously amplified product, before a 10-minute incubation at 95 °C inactivates the UNG prior to amplification. Since UNG treatment and subsequent amplification is performed within a closed tube, there is no potential for further contamination.

Table 6.11 Range of post-PCR analysis approaches.

Technique	Analysis Method	Detection method	Comments	Advantages	Disadvantages
Agarose gel electrophoresis	PCR products are separated by size using agarose gel electrophoresis, and sized by comparison to a DNA ladder of known molecular weight fragments run on the same gel	Usually by use of an intercalating dye such as ethidium bromide (Limit of detection in ng range)	Resolution of products is determined by % agarose used[54]	Relatively cheap and quick	Non-specific as will detect any nucleic acid present
		SYBR® Green I (Limit of detection in pg range)	Higher resolution agaroses available such as NuSieve (FMC Bioproducts) MetaPhor (FMC Bioproducts)	Easy to set up	Stains often carcinogenic, although safer dyes now available (for example SYBRsafe) Low throughput Qualitative method, some gel documentation systems offer semi-quantitative methods
Acrylamide gel electrophoresis	Electrophoresis separates products by size, which is determined using a molecular weight marker, run on the same gel	Intercalating dyes	Resolution of products is determined by % acrylamide used[54]	Higher resolution than agarose gels	More difficult to handle than agarose gel and more laborious
		Silver staining		Semi-quantitative	Unpolymerised acrylamide is neurotoxic
		(Limit of detection in pg range)		Sequencing gels confirm specificity	Sequencing equipment expensive

(*Continued*)

Table 6.11 (Continued)

Technique	Analysis Method	Detection method	Comments	Advantages	Disadvantages
		Fluorescent detection using labelled nucleotides			High voltages required
Solid Phase Capture	Capture of amplified fragments of interest onto a solid support such as Dynabeads®	Detection of the bound product may be through use of primer labelled with enzyme such as alkaline phosphatase or horseradish peroxidase, or with a directly labelled fluorescent molecule such as fluorescein		Overcome the problems associated with electrophoresis Automated and therefore high throughput	Long developmental time required and extensive evaluation and validation before routine use
Capillary electrophoresis	Separation takes place in a thin fused silica-coated capillary which contains the matrix Separation again is based on size differences of products	On line detection by UV absorbance or fluorescent detection	Traditionally required highly skilled staff to run these instruments However new 'plug and play' instruments and software are available such as	Rapid analysis of small molecules Sensitive: UV detection in pg–fg range. LIF detection in amol–zmol range	Equipment more expensive

MALDI-ToF MS	Matrix-assisted laser desorption-ionisation is used to ionise PCR products or other oligonucleotide targets such as single base extension (SBE) products. These are then accelerated by an electric field and distinguished by the time taken to reach the detector. The time of flight (ToF) measurement is then converted to mass	As the ions reach the detector surface the charge induced or current produced is recorded (as mass to charge or m/z ratios). The intensity of the signal is proportional to the number of molecules, permitting quantification	Soft ionisation techniques such as electrospray ionisation (ESI) and MALDI have enabled analysis of DNA molecules without fragmentation	the Bioanalyzer 2100[55] High sample throughput New instruments easy to use Quantitative data Calibrant standards of known mass run daily or per batch Individual standards not required as instrument is then calibrated to convert ToF to mass Rapid, very high accuracy measurement possible Can be configured for high-throughput applications	Expert machine operation required Instrumentation expensive Extensive and stringent sample purification required Size of products that can be analysed is limited, so suited to primer extension-type assays or analysis of fragmented or cleaved molecules

6.6.3.2 Ultraviolet Light

The use of ultraviolet light has been recommended for the decontamination of surfaces[44] and for reagent solutions[47–49] and acts by dimerising neighbouring pyrimidines (CT, TT, TC, CC) thereby inhibiting polymerisation. In practice, the effect is reversible as irradiated targets exist in an equilibrium state, with the consequence that short PCR products may be relatively UV resistant depending upon their sequence.[50] Achieving sufficient levels of exposure to UV is problematic where surfaces are not perpendicular to the light source, such as three-dimensional objects.

6.6.3.3 Chemical Decontamination

Alternatives to UV involve chemical decontamination, which may be more applicable to three-dimensional surfaces, depending upon their chemical resistance. A one-minute wash with 10% Clorox®, which is relatively non-corrosive, is sufficient to prevent PCR contamination.[51]

6.7 Post-PCR Analysis

There are a number of methods available for detecting PCR products, some of which are outlined in Table 6.11. The techniques mainly rely on distinguishing PCR products according to size, with either fluorescent detection, UV absorption or electrical detection through mass spectrometry. Alternative methods, such as electrical and electrochemical detection of hybridisation of PCR products,[52,53] are being developed.

6.8 Conclusions

It is clear from the considerations covered in this chapter that there are many variables that may affect the outcome, reproducibility and reliability of PCR. Initial care in design and optimisation of an assay can save time spent in rectifying analytical problems later. In addition, use of appropriate controls in PCR assays can identify changes in performance, or problems such as contamination at an early stage, enabling corrective measures to be taken to overcome them.

Ideally controls should be routinely included in experimental designs, and also performed realistically. For example, it is of little use to include negative controls in an experiment if the reactions are set up and closed before any chance of contamination has occurred. Similarly positive controls should be comparable to the actual unknowns in an assay, in terms of target concentration range and quality of target DNA. Treating controls in the same way as analytical samples can then give a high degree of confidence in the results, and allow meaningful interpretation.

Most PCR is performed in a shared environment, and therefore it is important that everyone working in the laboratory is aware of the precautions to be taken to minimise contamination. Taking responsibility for ensuring that DNA-containing waste materials are cleared away regularly, that working surfaces are cleaned before and after use and that pipettes are checked regularly and stored appropriately can make a significant difference to the outcome of experiments in the PCR laboratory.

References

1. R. K. Saiki, S. Scharf, F. Faloona, K. B. Mullis, G. T. Horn, H. A. Erlich and N. Arnheim, *Science*, 1985, **230**, 1350.
2. R. K. Saiki, D. H. Gelfand, S. Stoffe, S. J. Scharf, R. Higuchi, G. T. Horn, K. B. Mullis and H. A. Erlich, *Science*, 1988, **239**, 487.
3. L. Settanni, A. Corsetti, *J. Microbiol. Methods*, 2007, **69**, 1.
4. A. Holst-Jensen, S. B. Ronning, A. Lovseth and K. G. Berdal, *Anal. Bioanal. Chem.*, 2003, **375**, 985.
5. S. Yang and R. E. Rothman, *Lancet Infect. Dis.*, 2004, **4**, 337.
6. V. Onofri, F. Alessandrini, C. Turchi, M. Pesaresi, L. Buscemi and A. Tagliabracci, *Forensic Sci. Int.*, 2006, **157**, 23.
7. Y. M. Lo and K. C. Chan, *Methods Mol. Biol.*, 2006, **336**, 1.
8. M. L. Altshuler, *PCR Troubleshooting: The Essential Guide*, Caister Academic Press Publication, Moscow Research Institute of Medical Ecology, 2006, ISBN-13 978-1-904-455073.
9. C. R. Newton and A. Graham (ed.), *PCR (Introduction to Biotechniques, Second edition)*, Promega, Madison, WI, 1997, ISBN-13 978-0-387-915067.
10. T. Weissensteiner, H. G. Griffin and A. M. Griffin, *PCR Technology: Current Innovations, Second Edition*, CRC Press Inc., Boca Raton, FL, 2003, ISBN-13 978-0-849-311840.
11. K. B. Mullis, F. Ferre and R. A. Gibbs, *The Polymerase Chain Reaction*, Birkhauser Boston, 1994, ISBN-13 978-0-817-637507.
12. M. A. Innis, D. H. Gelfand and J. J. Sninsky, *PCR Strategies*, Academic Press, San Diego, CA, 1995, ISBN-13 978-0-123-721839.
13. P. Frame,http://www.mbi.ufl.edu/~rowland/protocols/pcr.htm (accessed May 2007).
14. http://www.dnastar.com/ (accessed May 2007).
15. https://products.appliedbiosystems.com/ (accessed May 2007).
16. S. J. Scharf, G. T. Horn and H. A. Erlich, *Science*, 1986, **233**, 1076.
17. P. D. Siebert and J. W. Larrick, *Biotechniques*, 1993, **14**, 244.
18. O. Landt, H. P. Grunert and U. Hahn, *Gene*, 1990, **96**, 125.
19. E. S. Stoflet, D. D. Koeberl, G. Sarkar and S. S. Sommer, *Science*, 1988, **239**, 491.
20. K. C. Kain, P. A. Orlandi and D. E. Lanar, *Biotechniques*, 1991, **10**, 366.
21. K. S. Lundberg, D. D. Shoemaker, M. W. Adams, J. M. Short, J. A. Sorge and E. J. Mathur, *Gene*, 1991, **108**, 1.
22. http://www.stratagene.com/ (accessed May 2007).

23. http://www.stratagene.com/newsletter/pdf/13_2_p65.pdf (Improve amplification specificity with Hotstart PCR enzyme) (accessed May 2007).
24. http://www.promega.com/ (accessed May 2007).
25. http://www.clontech.com/ (accessed May 2007).
26. http://www.sigmaaldrich.com/ (accessed May 2007).
27. http://www.invitrogen.com/ (accessed May 2007).
28. B. D. Cobb and J. M. Clarkson, *Nucleic Acids Res.*, 1994, **22**, 3801.
29. G. Taguchi, *Introduction of Quality Engineering*, Asian productivity Organisation, UNIPUB, NY, 1986.
30. G. Taguchi and Y. Wu, *Introduction to Offline Quality Control*, Japan Quality Control Organisation, 1980.
31. L. Rossen, P. Norskov, K. Holmstrom and O. F. Rasmussen, *Int. J. Food Microbiol.*, 1992, **17**, 37.
32. D. H. Gelfand, in *PCR Technology, Principles and Applications for DNA Amplification*, ed. H. A. Erlich, Stockton Press, New York, NY, 1989, ISBN-13 978-0-195-098754, p. 17.
33. W. Abu Al-Soud and P. Radstrom, *Appl. Environ. Microbiol.*, 1998, **64**, 3748.
34. D. L. Wiedbrauk, J. C. Werner and A. M. Drevon, *J. Clin. Microbiol.*, 1995, **33**, 2643.
35. H. A. Powell, C. M. Gooding, S. D. Garrett, B. M. Lund and R. A. McKee, *Lett. Appl. Microbiol.*, 1994, **18**, 59.
36. L. Herman and H. De Ridder, *Neth. Milk Dairy J.*, 1993, **47**, 23.
37. J. Bickley, J. K. Short, D. G. McDowell and H. C. Parkes, *Lett. Appl. Microbiol.*, 1996, **22**, 153.
38. http://www.epibio.com/item.asp?id=294 (accessed May 2007).
39. T. A. Patterson, E. K. Lobenhofer, S. B. Fulmer-Smentek, P. J. Collins, T. M. Chu, W. Bao, H. Fang, E. S. Kawasaki, J. Hager, I. R. Tikhonova, S. J. Walker, L. Zhang, P. Hurban, F. de Longueville, J. C. Fuscoe, W. Tong, L. Shia and R. D. Wolfinger, *Nat. Biotechnol.*, 2006, **24**, 1140.
40. http://www.microzone.co.uk/products/ok.html (accessed May 2007).
41. http://www.roche-applied-science.com/lightcycler-online/ (accessed May 2007).
42. http://www.bio-rad.com/iCycler (accessed May 2007).
43. http://www.corbettlifescience.com/ (accessed May 2007).
44. S. Kwok and R. Higuchi, *Nature*, 1989, **339**, 237.
45. K. H. Rand and H. Houck, *Mol. Cell Probes*, 1990, **4**, 445.
46. M. C. Longo, M. S. Berninger and J. L. Hartley, *Gene*, 1990, **93**, 125.
47. G. Sarkar and S. S. Sommer, *Methods Enzymol.*, 1993, **218**, 381.
48. S. T. Isaacs, J. W. Tessman, K. C. Metchette, J. E. Hearst and G. D. Cimino, *Nucleic Acids Res.*, 1991, **19**, 109.
49. A. Meier, D. H. Persing, M. Finken and E. C. Bottger, *J. Clin. Microbiol.*, 1993, **31**, 646.
50. R. W. Cone and M. R. Fairfax, *PCR Methods Appl.*, 1993, **3**, S15.
51. A. M. Prince and L. Andrus, *Biotechniques*, 1992, **12**, 358.

52. L. Moreno-Hagelsieb, B. Foultier, G. Laurent, R. Pampin, J. Remacle, J. P. Raskin and D. Flandre, *Biosens. Bioelectron.*, 2006, **22**, 2199.
53. T. Nojima, K. Yamashita, A. Takagi, Y. Ikeda, H. Kondo and S. Takenaka, *Anal. Sci.*, 2005, **21**, 1437.
54. J. Sambrook, E. F. Fritsch and T. Maniatis (ed.), *Molecular Cloning – A Laboratory Manual*, Cold Spring Harbor Laboratory Press, US, 1997, ISBN-13 978-0-879-693091.
55. http://www.home.agilent.com/ (accessed May 2007).

CHAPTER 7
Quantitative Real-time PCR Analysis

JACQUIE T. KEER

LGC, Queens Road, Teddington, TW11 0LY

7.1 Introduction

The sensitivity of analysis achievable with PCR has led to the technology being adopted across a range of sectors. For many applications a quantitative result is required, which has driven the development of a range of strategies to determine the amount of starting material in a sample. Approaches such as competitive PCR[1] and limiting dilution analysis[2] have been used as routes to quantification, although the variable nature of the PCR process and the amplification of the target to a maximal level irrespective of the starting amount of target limit the accuracy of these methods.[3]

The advent of kinetic or real-time PCR[4] has overcome many of the limitations of earlier strategies, by monitoring the increase in product generated during the course of the reaction, in 'real time'. Quantitative approaches are based on the time or cycle at which amplification is first detected, rather than requiring quantification of PCR products, and the principle is illustrated schematically in Figure 7.1. A range of samples of known target content are usually amplified together with the samples under test, and the accumulation of PCR product in each cycle is determined. Alternatively the signal from two targets may be compared to determine a relative measure of quantification, and this is often used in measurement of gene expression which is considered in more detail in Chapter 9.

Here a fluorescent reporter assay is used to monitor increase in fluorescence at each PCR cycle. The point at which the signal becomes detectable, or crosses some arbitrary threshold value, is determined for each standard and sample. These values are then plotted against the amount of target in the standards to produce a calibration curve, and the amount of target in the unknown samples can then be interpolated from the graph.[5]

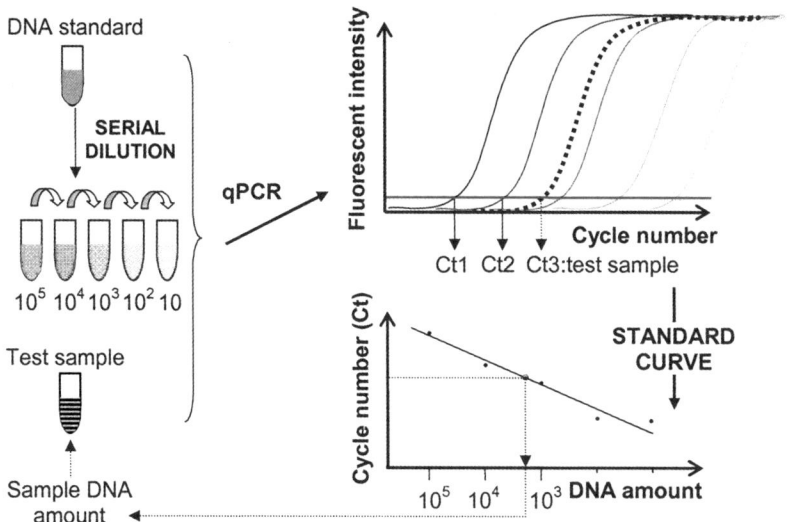

Figure 7.1 Schematic representation of the basis of real-time PCR using a quantitative standard.

The linear relationship between the amount of starting material and the measured cycle threshold (Ct) values are maintained across several orders of magnitude, so assays based on quantitative PCR (qPCR) have an unusually large dynamic range. There are a number of other significant benefits in using real-time PCR analysis, including the greatly increased sensitivity associated with the use of fluorescent reporters and signal collection devices, and the rapid cycling times that are achievable on some instruments. In addition, homogeneous qPCR assays minimise the potential for cross-contamination compared with conventional methods as reaction vessels need not be opened in order to analyse amplification products, and also avoid variation introduced by gel analysis.

In short, real-time PCR offers the potential of well-characterised and highly sensitive quantitative analysis, although the diversity of instruments, detection chemistries, data handling methods and the lack of quantitative reference standards present significant challenges to measurement comparability.

7.2 Approaches to Product Detection

The key feature of quantitative real-time PCR is that the amount of product is measured at each cycle of the reaction, and thus requires simultaneous PCR amplification and product detection. The first assay of this type utilised simple incorporation of the fluorescent dye ethidium bromide into the PCR reaction, and the increase in signal resulting from the dye intercalating into the double-stranded PCR products was monitored using a CCD camera.[4] However, ethidium bromide signals will increase with increasing amounts of any double-stranded DNA, and thus primer dimers and non-specific PCR products will all generate signals that are not distinguishable from a true positive reaction, therefore more specific reporter systems are preferable for many applications. In addition, more efficient fluorophores than ethidium bromide are available (Chapter 5), which have been successfully employed to improve the sensitivity of real-time assays. The majority of homogenous assays rely on the transfer of energy between fluorescent reporter and quencher molecules to generate specific signals. When oligonucleotide probes are excited the energy absorbed by the fluorophore may be emitted as fluorescence or may be transferred to a quencher and released as heat or light of a different wavelength. This energy transfer may occur through Förster Resonance Energy Transfer (FRET) if the emission and absorption spectra of the molecules overlap sufficiently or through non-FRET mechanisms by short-range contacts, which do not require any spectral overlap between the donor and acceptor molecules. Several of the most common approaches that have been developed to monitor PCR kinetics, and the labels used, will be described in the following sections.

7.2.1 The 5' Nuclease Assay

The 5' nuclease, or TaqMan™, assay[6] utilises FRET quenching to analyse PCR-amplified target DNA, although the original method was developed using a radioactive labelling approach.[7] The assay exploits the 5'-3' nuclease activity of *Taq* DNA polymerase to cleave a dual-labelled oligonucleotide probe, labelled with fluorophore and quencher moieties on the 5' and 3' termini respectively. Little fluorescence is emitted from the intact probe due to efficient intramolecular quenching, as energy absorbed by the fluorophore is transferred to the quencher and dissipated as heat. However, during PCR amplification, TaqMan™ probes specifically hybridise to their target sequences and the 5'-3' exonuclease activity of *Taq* polymerase cleaves the probes between fluorophore and quencher moieties. Enzymatic cleavage of TaqMan™ probes spatially separates fluorophore and quencher components, causing significant increase in fluorescence emission (Figure 7.2). With each cycle of denaturation, primer annealing and product extension, a molecule of reporter dye is liberated from a quencher moiety for each molecule of newly synthesised DNA. Therefore, the magnitude of the emission increase produced during amplification is

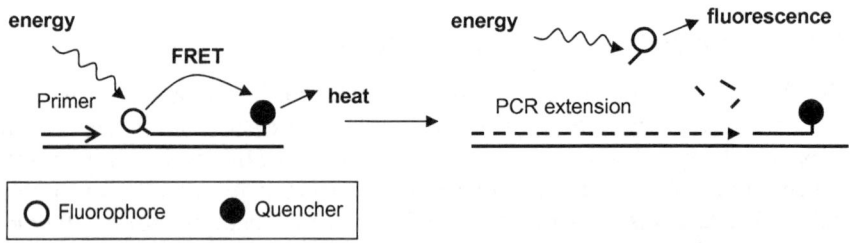

Figure 7.2 Illustration of the basis of the 5' nuclease (TaqMan™) assay.

proportional to the amount of PCR product synthesised. Fluorescence emission may be monitored throughout the course of PCR, allowing the generation of a 'real-time' representation of target amplification. Polynucleotide targets differing by as little as a single nucleotide may be distinguished using TaqMan™ probes, since oligonucleotides hybridise to mismatched DNA targets with a significantly reduced efficiency. Careful design of TaqMan™ probes allows discrimination of polymorphic targets, where only perfectly matched probes are degraded during amplification generating increases in fluorescent signal.[6,8] TaqMan™ probes may be employed to detect and discriminate multiple targets in a single reaction, using oligonucleotides labelled with spectrally distinct fluorophores,[9] and a wide range of commercial assays have been developed. Specific software is available to facilitate the design of 5' nuclease assays, and up-to-date guidelines can also be found on the Applied Biosystems website.[10] In brief, the melting temperature of the probe should be 10 °C higher than that of the primers, and should be located adjacent to one of the primers but not overlapping the primer binding site. The inclusion of a G at the 5' end of the probe should be avoided, as this base may partially quench the signal from the fluorophore. As the requirement for a probe of much higher T_m than the primers can be a challenge in designing probes, molecules that are designed to bind to the minor groove of dsDNA can be employed (MGB probes) to achieve the required T_m without using lengthy sequences.[11] A similar improvement in probe binding and T_m can be achieved by using locked nucleic acid probes, LNAs,[12] which are RNA analogues with a structurally constrained sugar-phosphate backbone.

7.2.2 Molecular Beacons™

Molecular Beacons™ are essentially single-stranded oligonucleotide probes that are non-fluorescent in isolation, but become fluorescent upon hybridisation to target sequences.[13,14] Non-hybridised molecular beacons form stem-loop structures, possessing a fluorophore covalently linked to one end of the molecule and a quencher linked to the other, such that the hairpin of the beacon places the fluorophore moiety in close proximity with the quencher. Since the quencher component is commonly a non-fluorescent moiety, the energy it receives from the fluorophore is released as heat, such that fluorescence is not emitted from unhybridised probe. The loop portion of the molecular beacon molecule is a specific probe that is complementary to a nucleic acid sequence present in the target DNA. Probe-target duplexes are longer and more stable than the stem hybrids. Therefore, when molecular beacons hybridise to target sequences, they undergo a fluorogenic conformational change where fluorophore and quencher moieties become spatially separated, such that the fluorophore is no longer quenched and the molecular beacon fluoresces (Figure 7.3). In designing molecular beacons, a sequence between 10 and 40 nucleotides long is chosen in the centre of the PCR product, which is complementary to the target of interest. The sequence should be free of significant secondary structure, then the stem and loop structures are formed by adding 5–7 bases and a fluorophore

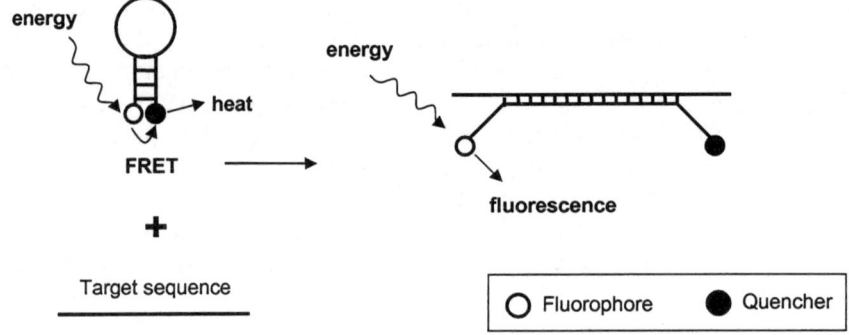

Figure 7.3 Target detection using Molecular Beacons™.

at the 5' end and a complementary 5–7 nucleotides and a quencher at the other terminus.

The secondary structure of the molecular beacon conveys high specificity to the probe, allowing the identification of targets that differ by a single nucleotide, where only perfectly complementary probe-target duplexes are sufficiently stable to induce the fluorogenic conformation transition. Molecular Beacons™ may also be employed to detect and discriminate multiple targets in a single reaction, using probes that possess different fluorophores which emit light at spectrally distinct wavelengths[15] (Table 7.1).

Simpler linear probe systems have been developed, which only require a single labelled reporter. HyBeacons® have a single fluorescent label attached to an internal nucleotide, and show enhanced fluorescence on binding to a complementary target.[16] The ResonSense® system is based on FRET, but an intercalating fluorophore is used as the donor, and a single label acceptor is attached to either the 3' or 5' end of the probe or internally.[17] These simpler reporters are well suited to multiplexed assays.

7.2.3 Hybridisation Probes

Hybridisation probes are oligonucleotides that are singly labelled with a fluorophore. Two such oligonucleotides are required for each hybridisation probe assay, one labelled with a donor fluorophore and the other with an acceptor fluorophore.[18] Excitation of the donor fluorophore produces an emission spectrum that overlaps with the absorption spectrum of the acceptor fluorophore. Hybridisation probe pairs are designed to recognise adjacent nucleotide sequences within target molecules. In isolation, the acceptor oligonucleotide is not excited and does not generate a fluorescent signal. However, during hybridisation to the target sequences, the donor and acceptor probes are brought into close proximity, allowing fluorescence resonance energy transfer from the donor to the acceptor (Figure 7.4).

Fluorescent signal from the acceptor fluorophore is thus emitted when both probes are hybridised to the target molecule. When incorporated into PCR

Table 7.1 Primary λmax absorption and emission wavelengths of frequently used fluorophores, quenchers and fluorescent dyes.

Name	Absorption max (nm)	Emission max (nm)
Fluorophores		
EDANS	336	490
FAM	492	515
Fluorescein	494	525
SYBR Green I	497	520
Ethidium bromide	518	605
JOE	520	548
TET	521	536
Yakima Yellow	525	548
VIC	528	546
HEX	535	556
Cy3	544	570
TAMRA	555	580
ROX	575	602
Texas Red	583	603
LC-RED 640	625	640
LC-RED 705	625	603
Cy5	647	667
Quenchers		
DABCYL	471 (∼400–550)	—
Deep Dark Quencher I	410 (∼400–550)	—
Deep Dark Quencher II	630	
Eclipse	530	—
Black Hole Quencher-1	534 (∼480–580)	—
Black Hole Quencher-2	579 (∼559–650)	
Black Hole Quencher-3	672 (620–730)	
Iowa Black FQ	532	—
Iowa Black RQ	645	
QSY-7	571	—
QSY-21	660	

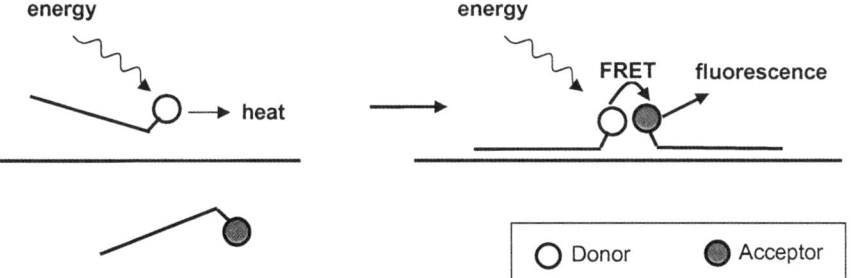

Figure 7.4 Target detection using hybridisation probes.

reactions, fluorescence from the acceptor probe is monitored once per cycle of amplification, allowing real-time measurement of product accumulation, where the amount of fluorescence emitted by the acceptor is proportional to the quantity of target synthesised. The use of several oligonucleotide probes bearing spectrally distinct acceptor fluorophores may be employed in a multiplexed analysis, to simultaneously detect and discriminate multiple targets in a single PCR reaction[19] (Table 7.1).

7.2.4 Scorpion™ Primers

In contrast to the systems already described, where the probe and PCR primers are located on separate DNA strands, Scorpion™ primers are designed so that the probe, primer and hence the amplified target are located on the same DNA molecule. Scorpion™ probes comprise a primer with an attached probe tail sequence, where the probe is contained within a stem-loop secondary structure similar to that of a molecular beacon.[20] In the unextended form, Scorpion™ primers are non-fluorescent due to fluorophore and quencher moieties being in close proximity. During PCR, the primer component of the Scorpion™ is extended at its 3' end producing the homologous target sequence required for probe hybridisation. When the Scorpion™ probe sequence hybridises to the amplified target, the hairpin loop of the probe opens and the fluorophore and quencher moieties become spatially separated (Figure 7.5) causing significant increases in fluorescent emission. The fluorescent signal is produced concurrent with target amplification, allowing the amount of product to be monitored. A benefit of this system is that unimolecular binding events are kinetically favoured over bimolecular hybridisation.

7.2.5 Plexor™ Primer Technology

Plexor™ primer technology, available from Promega, is a relatively new approach that requires only two primers for sensitive and specific quantification of amplified DNA. The approach exploits the highly specific interaction between structurally modified G and C bases, isoguanine (iso-dG) and 5'-methylcytosine (iso-dC) respectively, which only base-pair with each other when incorporated into dsDNA.[21] The approach involves synthesising one of

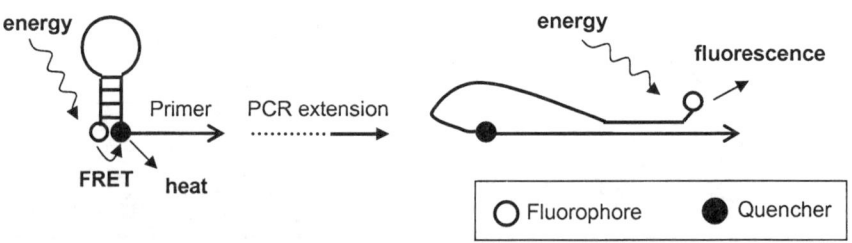

Figure 7.5 Target detection using a Scorpion™ probe.

the 2 PCR primers with an iso-dC base and a fluorophore at the 5′ end, and including iso-dG bases modified to include a quencher in the PCR reaction. As the amplification progresses, only modified iso-dG can be incorporated into products complementary to the primer, and thus the quencher is brought into close enough proximity to the fluorophore to effect quenching of the signal, at a level proportional to the amount of PCR product generated. An added benefit of this system is that the quenching is reversible, so melt curve analysis can be performed using this system.

7.2.6 Melting Curve Analysis

In addition to measuring the increase in product at each cycle, an analysis of the products generated in the reaction may be performed at the end of the amplification process. This is termed 'melt analysis' and is compatible with both intercalating dye reporter systems and those where the probe binds to the PCR product to achieve a change in fluorescence intensity. To perform the analysis the fluorescent signal is monitored as the temperature is gradually increased from around 50 to 95 °C, which results in an increase in fluorescence as the dsDNA or probe:product complex is dissociated. The change in fluorescence against temperature is usually plotted by instrument software, yielding peaks corresponding to the denaturation maxima of each double-stranded species present in the system (Figure 7.6). Primer-dimers, non-specific or mismatched sequences will generally have a lower melting temperature than the specific product of the reaction, and so can be distinguished by this post-PCR analysis.[22]

Recently, an extension of melt curve analysis has been developed, termed High Resolution Melt or HRM (Corbett Life Science), enabled by improvements in

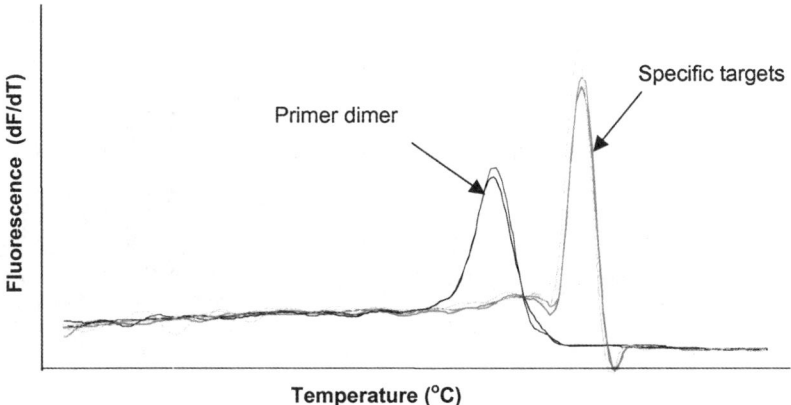

Figure 7.6 Schematic representation of a typical post-amplification melt analysis to differentiate specific PCR products from primer dimers. Melt analysis can also be used to identify wild type and mutant alleles, based on the lower melting temperature of unmatched species.

real-time instrument capabilities and the dyes used in analysis. Rapid data collection is required, with very high precision thermal resolution (down to 0.02 °C) and dedicated analysis software.

Samples are characterised based on very detailed measurement of their disassociation kinetics, and with more detailed melt profiles samples may be discriminated by length, GC content and sequence. Even single base-pair mismatches can be distinguished, allowing application of the method to detailed genotyping analysis, and thus this approach could potentially be exploited to replace the use of more complex probe reporter systems.[23]

7.2.7 Choice of Fluorophores

A range of dyes is available from various manufacturers, which may be used in constructing probes for use in the assays described here. Considerations in choosing a dye include the type of assay design, the limitation of the instrument in terms of excitation and signal capture wavelengths, and the overlap of fluorophore signals if multiplexing is required. The majority of fluorophores have wide emission spectra, and there is often a significant degree of overlap in signals (Figure 7.7). The labels used in a multiplexed assay must be chosen

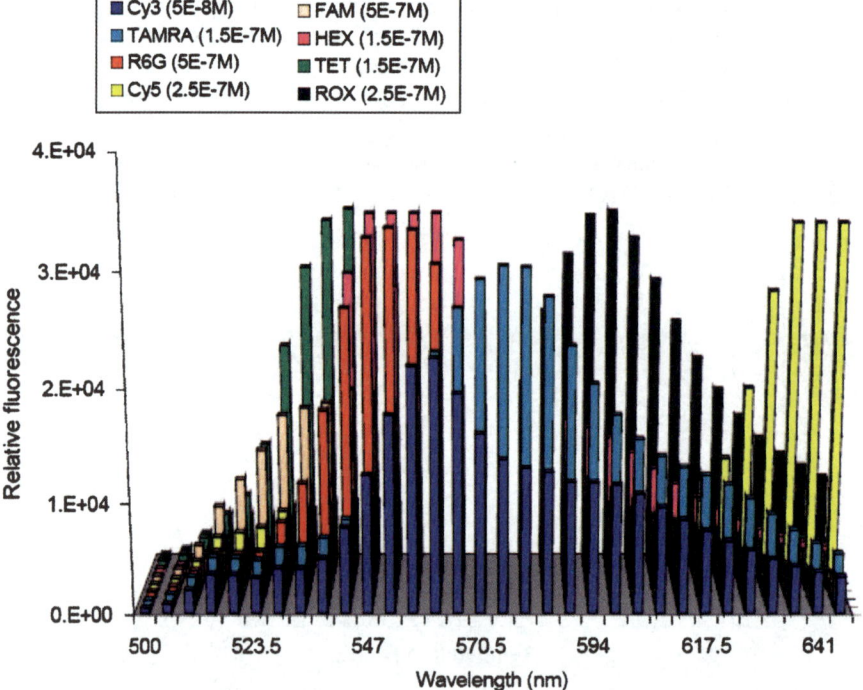

Figure 7.7 Spectra of a number of commonly used fluorophores measured using the ABI PRISM® 7700 SDS. The molar concentrations of each fluorophore are indicated in the key.

carefully to maximise spectral separation, and thus facilitate deconvolution of the multiplexed signals.

Usually assays are designed to utilise a universal donor molecule, which absorbs energy from the instrument light source and transfers it to a variety of adjacent reporter molecules. The signal from the reporters is distinguishable by the emission wavelengths, which are measured by the instrument. The ResonSense® assay is an example of this design, where the intercalating fluorophore is excited and transfers energy to a range of labelled probes. The alternative approach is to have a range of donor molecules which all may be excited by the light energy from the instrument, whilst a universal acceptor molecule is utilised as a quencher. The 'TaqMan™' assay run on the ABI real-time platforms is an example of the second type of assay, where TAMRA or an alternative quencher is paired with a variety of 5' fluorophore labels.

In designing FRET-based assays there should be sufficient spectral overlap between the fluorophore-quencher pairs, such that the emission maximum of the donor is within the excitation range of the acceptor molecule. For example, fluorophores with an emission maximum between 500 and 550 nm, such as FAM and TET, are effectively quenched by dyes such as DABCYL (absorption maxima at 471 nm, with a range between 400 and 550 nm). By contrast, fluorophores with a higher emission maximum such as ROX, Texas Red and the Cy dyes are more effectively paired with quenchers with a higher absorption maximum (Table 7.1).

7.3 Range of Instruments

The number of real-time PCR instruments on the market is still increasing, offering a variety of options in terms of throughput, cycling times, flexibility and cost[10,24–30] (Table 7.2). The basic requirements of an instrument are to provide the temperature-controlled environment for PCR amplification, whilst providing light excitation and quantitative fluorescent signal collection of appropriate wavelength; thus both controlling and monitoring the amplification process. Instruments differ in many key features, including:

- Speed of reaction (heating and cooling rates);
- Precision and uniformity of the temperature control;
- Throughput/number of reactions performed at one time;
- Range of excitation wavelengths;
- Range of detection wavelengths;
- Sensitivity of detection device;
- Ease of use and capabilities of software;
- Flexibility of temperature profiles and chemistries within each run;
- Cost.

The choice of instrument will depend primarily on the range of applications for which it is required.

Table 7.2 Performance characteristics of a selection of commercially available real-time PCR instruments.

Instrument	ABI PRISM® 7900HT SDS	ABI StepOne®	Stratagene Mx® 3000P® (4000) Systems	Corbett Rotor-Gene™ 6000	Roche Lightcycler® 480	Roche LightCycler® 2.0	Eppendorf MasterCycler®	Cepheid Smart Cycler®	BioGene InSyte
Capacity	96 or 384	48-well block	96-well block	36 or 72 tubes	96 or 384	32 capillary	96-well block	16 to 96 tube	96 reactions
Reaction volume/μl	Standard: 25–100; Fast: 10–30; 384: 5–20	10–30	20–100, optimised for 25	5–100	96: 20–100; 384: 5–20	20 and 100	25–100	2–100	50
Reaction time (40 cycles)	Standard: 2 h; Fast: 33 min (96), 52 min (384)	40 mins/2 hours	2.5 hours (1.5 h)	30 minutes	40 min (384); 60 min (96)	30 min (30 cycles)	30 min	40 min	<20 min (35 cycles)
Excitation source and range	488 nm argon-ion laser	Single blue LED	Quartz Tungsten Halogen lamp 350 nm–750 nm	4 separate high-power LEDs; 470, 530, 585 and 625 nm	High-intensity xenon lamp 430–630 nm and 5 filters: 450, 483, 523, 558, 615 nm	Blue LED 470 nm	96 LEDs at 470 nm	4 channels: 450–495, 500–550, 565–590 and 630–650 nm	Solid-state blue laser 473 nm
Detector/Emission range	Spectrograph and cooled CCD	Three emission filters, and photodiode	1 (4) scanning photomultiplier tubes (PMT),	PMT and 6 detection filters; 510, 555, 610,	CCD camera with 6 detection	6 channel photometer: 530, 560,	1- or 2-channel PMT; 520/550	4 channels; 510–527, 565–590,	32 channel spectrometer (520–720 nm)

	camera from 500-660 nm		with 4 filters; 350-830 nm	665, 570 and 610 nm	filters; 500, 533, 568, 610, 640 and 670 nm	610, 640, 670 and 705 nm	or 520/550/580/605 nm	606-650 and 670-750	
Temperature control	Peltier	Peltier	Peltier hybrid (resistive + convective)	Air heating and cooling, with centrifugation	Peltier with Therma-base heat distribution	Air heating and cooling	Normal and fast Impulse PCR blocks	Solid-state heater and forced air cooling	Electrically conducting polymer (ECP)
Temp. ramp rate °C/sec	Standard: ±1.6 °C/sec; Fast: ±3.0 °C/sec	Standard ±1.6 °C/sec; Fast: ±2.2 °C/sec	Up to 2.5 °C/sec	10 °C/sec		20 °C/sec	Fast: 6 °C/sec heating, cooling 4.5 °C/s	Heating, 10 °C/s; cooling 2.5 °C/s	Av. 15 °C/sec
Temp. uniformity	±0.50 °C	±0.50 °C	±0.25 °C	±0.01 °C	±0.1 °C	±0.3 °C	35 °C ±0.3 °C, 90 °C ±0.4 °C	±0.50 °C	0.01 °C resolution
Temp. range	4 °C-100 °C	4 °C-100 °C	25-95 °C	25-99 °C	40-95 °C	40-98 °C	4 °C-99 °C	50-99 °C	Ambient - 100 °C
W x D x H/cm	72 × 64 × 84	24.6 × 51 × 42.7	33 × 46 × 43	38 × 48 × 31.5	60 × 60 × 54.5	28 × 38.5 × 50.5	26 × 41 × 39.6	30 × 30 × 25	44 × 46 × 60
Weight/kg	82	23.6	20	17	55	22	24	10	45
Additional features	Auto loader (throughput 5000 wells/8h day)			Optical temperature validation system				Independently controlled reactions	

Instruments generally fall into two types; higher-throughput machines with 96 or 384 reaction capacity for processing batches of samples (for example the ABI PRISM® 7900HT and the Roche LC480) or more flexible instruments with faster reaction times and greater flexibility, such as the Cepheid Smart-Cycler®.

Instruments using a single wavelength excitation source are more limited in the variety of compatible fluorophores than those with a broader excitation range. For example the Eppendorf MasterCycler® utilises LED excitation at 470 nm, while Stratagene' Mx3000P® employs a halogen lamp with an excitation range from 350–750 nm. The fluorescence detector and the capability of the instrument for distinguishing different wavelength signals also influences the level of assay multiplexing that is achievable.

Uniform and precise temperature control is central to obtaining reproducible quantitative results, and thus there is a requirement for instruments to have effective thermal control systems. Many of the instruments on the market utilise Peltier-driven heating and cooling systems, which work by passing electric currents through semiconductor elements connected in series to effect temperature changes that are proportional to the currents applied. Exceptions are the Corbett Rotor-Gene™, which houses reaction tubes in a centrifugal rotor, and the Roche LightCycler® 2.0, which contains a rotor of glass capillaries, both relying on air heating and cooling in the reaction chamber to control sample temperature. High-speed centrifugation of samples in the Rotor-Gene™ ensures temperature homogeneity between samples, and air heating and cooling systems afford both instruments rapid reaction times. The latest versions of LightCycler® 2.0 software also enable the samples to be constantly rotated during amplification, thereby increasing thermal homogeneity. Newer electrically conducting polymer (ECP) heating technology is exploited in BioGene' InSyte real-time instrument, enabling precise individual tube thermal control and reduced run times. Miniaturisation has also seen the introduction of nanofluidic chips for qPCR analysis, which are biochips employing systems of integrated channels and valves to manipulate the reagents and house the amplification process.[31]

7.4 Practical Aspects of qPCR Analysis

A number of factors can affect the performance of qPCR, including the initial choice of target, probe and primer sequences, the concentration and type of reaction components, the thermal cycling conditions, the reaction vessels used and the preparation of the samples and any standards used. To ensure reliability of analytical results both the assay design and the reagents should be considered carefully.

7.4.1 Assay Design

A range of validated assays are available from a number of manufacturers, such as the TaqMan™ SNP Genotyping and Gene Expression Assays from Applied

Biosystems, and Invitrogen' D-LUX™ assays. The benefits of using pre-designed assays are that the reagents and methods are usually validated and quality-controlled, saving time and effort. However an appropriate assay may not be commercially available for many applications, and the cost of utilising off-the-shelf assays may be prohibitive. An online database of quantitative PCR primers and probes (QPPD) designed for human and mouse gene expression studies is also available on line.[32]

7.4.1.1 Target Sequence

In designing a new assay, choice of target sequence is typically the first consideration. The target ideally should not contain strong secondary structure as this can reduce the efficiency of oligonucleotide probe hybridisation, and this may be assessed using the Mfold[33] or similar structure prediction program, several of which are freely available online.[34] The target sequence should also be analysed for the existence of similar sequences that may interfere with the assay using a BLAST search[35] of sequence databases, and the assay region chosen to minimise any cross reactivity.

The amplicon size may affect assay efficiency and sensitivity; if too long the double-stranded PCR products may not denature efficiently at each cycle, and may preferentially re-hybridise in each cooling cycle before probe and primer sequences can bind. Typically amplicons of less than 150 bp are used, although amplicon size is not limited by these considerations in assays utilising fluorescent dyes or intercalating dyes.

7.4.1.2 Probe and Primer Design

Having chosen the target sequence, primers and probe sequences will be required, depending on the chosen assay format. Several manufacturers of real-time instrumentation and assay reagents provide software for this purpose, and a number of primer and reporter probe design packages are also freely accessible online, such as FastPCR[36] and AutoPrime.[37] For most applications, primers are designed to be fully complementary to template DNA sequences, and the basic considerations are similar to those for successful conventional PCR. Typically, primers should be designed to be 18–30 nucleotides in length to allow a reasonably high annealing temperature to be used during PCR. Primer pairs should be approximately the same length, should possess 40–60% GC content and should lack significant secondary structures or complementary regions. Regions of complementarity at the 3' end of primers should be minimised to reduce the potential for the formation of primer dimer during amplification.

The design requirements for the individual types of fluorescent probes will not be discussed here, but can be found in the cited literature,[7,8,13,15,19,20] and generally the probe T_m should be higher than that of the primers to ensure that the probe has bound before the primers hybridise and extension begins. If the application requires discrimination of closely related target sequences, then probes of between 15 and 30 nucleotides in length are recommended, as this gives a balance between forming sufficiently stable hybrids with target sequences, but retaining the sensitivity of melting temperatures to the presence of sequence mismatches. AT-rich target sequences may require probes that are greater than 30 nucleotides in length to form stable hybrids, and probes composed of peptide nucleic acid (PNA) or containing certain DNA base analogues (for example Propyne dC) may also be employed to form more stable interactions. Conversely, target sequences that are particularly GC rich may require probes that comprise fewer than 15 nucleotides for effective target discrimination, or alternatively DNA base analogues (such as N4 Ethyl dC) may be used to destabilise duplex interactions and lower the T_m of oligonucleotide probes.

The type of nucleotide mismatch that occurs within DNA duplexes strongly influences the stability of hybridisation. Mismatched interactions involving G, particularly G to T, are the least destabilising whilst interactions involving C are the most destabilising.[38] The position of the base mismatch, relative to the probe/target duplex, also influences the difference in stability between matched and mismatched interactions. Positions of mismatch located at duplex termini are significantly less destabilising than mismatches situated towards the centre of oligonucleotide probes,[39] thus allowing design of the probe position according to the specific application requirements.

7.4.2 PCR Master Mix

The qPCR reaction environment is usually provided by a master mix that includes buffer, dNTPs, thermostable polymerase and $MgCl_2$. Additional components such as ROX as a passive reference dye, and UNG with dUTP to prevent PCR product contamination, may also be included. Many instrument manufacturers provide reagents for use with particular instruments or assay/probe formats, although many reagents work well with a variety of assays and platforms. Use of commercial reagents affords benefits in terms of licence to perform qPCR, quality assurances and batch-to-batch consistency, although the cost is higher than reagents prepared in-house. Using complete systems from one manufacturer, from the assay design software through to the master mix and instrument settings, can reduce the number of factors requiring optimisation, thereby saving much time and effort.

7.4.2.1 Magnesium Chloride

The concentration of $MgCl_2$ is known to have an impact on both the specificity and the yield of PCR; insufficient Mg^{2+} results in poor yields as the polymerisation rate of *Taq* polymerase is low; however, if the level of Mg^{2+} is too high the specificity of the reaction is compromised. In contrast to conventional assays, homogenous assays require higher $MgCl_2$ concentrations of around 3–5 mM to achieve efficient target amplification and detection. The presence of $MgCl_2$ increases the rate of DNA hybridisation,[40] enabling efficient hybridisation during the rapid cycling conditions used in many instruments.

7.4.2.2 DNA Polymerase

The type of DNA polymerase employed in homogeneous PCR assays may affect the sensitivity and efficiency of target amplification, detection and discrimination. For example, certain enzymes such as *Z-Taq*™ (TaKaRa) exhibit higher processivities and rates of PCR extension than standard *Taq* polymerase. Increasing the speed of product synthesis may allow the reduction of PCR hold times and the overall duration of amplification, especially when combined with the fast temperature transition rates of rapid cycling instruments. GC rich target sequences may also require a high denaturation temperature, necessitating the use of polymerases such as Stoffel fragment and Deep Vent with increased thermal stability that support denaturation in excess of 95 °C.

The use of hot-start PCR may improve assay performance in applications where it is important to minimise the formation of primer-dimer and other non-specific PCR products, and has also been shown to improve the assays on the LightCycler®, possibly by reducing binding of the enzyme to the glass tube surfaces.[41]

The 5′-3′ exonuclease activity is vital for fluorescent signal generation and target detection in TaqMan™ style assays. Commercially available DNA polymerases have been demonstrated to generate variant amounts of fluorescent signal when employed in TaqMan™ assays performed on a LightCycler®[42], suggesting assay performance may be affected by the choice of enzyme. The majority of assays using TaqMan™ probes employ AmpliTaq Gold™ (ABI) for efficient 5′-3′ exonuclease activity and the large increases in fluorescence emission that it produces during amplification.

7.4.3 Cycling Conditions

The real-time platforms on the market are all supplied with detailed recommendations, and it is advisable to consult the manufacturer' literature for information on instrument operation and settings. A set of generalised assay conditions for homogenous assays using SYBR® Green on the LightCycler® 2.0 and a TaqMan™ assay on the ABI PRISM® 7900HT are shown in Table 7.3. Instruments capable of supporting rapid cycling protocols are increasingly available, although care should be taken to optimise the reaction and use appropriate reagents to maintain assay performance as rapid cycling may affect the sensitivity and precision of the assay.[43] The QuantiFast (Qiagen) system for rapid cycling utilises a buffer additive (Q-bond) to significantly reduce annealing, denaturation and extension times.

The choice of either two or three temperature cycles depends on the type of probe and instrument selected; TaqMan™ probe assays use a combined annealing and extension, molecular beacon assays use a three-stage reaction, while Scorpion™ and hybridisation probe assays may use either two or three temperature cycling. In real-time PCR most amplicons are typically short, so the requirement for the polymerase to extend at the optimum temperature is

Table 7.3 Typical reaction conditions for two common instrument and assay combinations.

	SYBR® Green in LightCycler® 2.0	TaqMan™ probe assay on ABI PRISM® 7900HT
Reaction	5–20 µl volume in glass capillaries	25–50 µl volume in 96 well plate
Master mix	1 × master mix, 1–5 mM $MgCl_2$, 1 mM dNTPs, 1 U DNA polymerase	1 × master mix, 3–10 mM $MgCl_2$, 1 mM dNTPs, 1–3 U DNA polymerase
Primers	0.5 µM each (0.1–0.8 typical range)	0.5 µM each (0.1–0.8 typical range)
Probe/dye	1:10 000 dilution SYBR® Green I	100–900 µM probe
Initial denaturation	95 °C 30 s	50 °C 2 min (UNG reaction) 95 °C 10 min (hot start)
Cycling	25–60 cycles	25–50 cycles
Denaturation	95 °C 0 s	95 °C 15 s
Primer annealing*	55 °C 5 s	60 °C 1 min
Extension	72 °C 10 s	
Data acquisition	End of extension step	During annealing/extension
Melt curve analysis	95 °C 0 s 35 °C 2 min 35 °C–95 °C at 0.1 °C/s	

* The primer annealing temperature used will be determined by the primer sequences used in the specific assay.

not absolute. Thus a two-step protocol is recommended in TaqMan™ probe assay, for example, as combining the annealing and extension stages into one step is quicker than the three steps (less time is taken during ramping between hold steps), and the relatively high annealing temperature ensures reaction specificity.

The melting temperature of oligonucleotide primers and probes determines the annealing temperature, or the combined annealing/extension temperature, at which fluorescence acquisition is performed. Assays that are designed to detect the presence of DNA sequences should possess annealing temperatures that are lower than the T_ms of the primers and probes. However, selection of annealing temperature is more complicated when target detection and discrimination is required, as ideally at the fluorescence acquisition temperature the oligonucleotide probe should be hybridised to perfectly matched sequences but should not be hybridised to mismatched targets, thus only generating signal from the fully complementary target. Optimisation of the annealing temperature is required to maximise the quantity of fluorescent signal emitted from matched probe whilst minimising the amount of mismatched probe that is hybridised. Achieving a reliable discriminatory assay using signal accumulation may not be possible if the matched and mismatched probe duplexes exhibit only small differences in T_m (less than 1–2 °C).

The hold times required in PCR cycles are determined by the type of polymerase, probe and instrument utilised. For example, if the Z-Taq™ enzyme and the rapid cycling conditions of the LightCycler® are used to amplify target sequences, denaturation, annealing and extension hold times may be reduced to 0 seconds, such that a 40 cycle amplification takes approximately 10 minutes to complete. The extension temperature of PCR may also affect the hold time required. At 72 °C, *Taq* polymerase adds significantly more nucleotides per second to extending products than at the 58–65 °C combined annealing/extension temperatures employed in two-stage PCR cycles. The hold time for PCR extension also depends on the size of the product being amplified, as larger targets require longer extension times.

One further consideration is to ensure that the excitation of the fluorophore is not too high, resulting in irreversible photobleaching and loss of fluorescent signal. Utilising less labile fluorophores, or performing the first ten cycles of amplification without fluorescence acquisition, can minimise loss of signal.

7.4.4 Primer and Probe Optimisation

A number of parameters will require optimisation in developing a robust assay, including the concentration of the primers, the amount of probe/intercalating dye reporter, the $MgCl_2$ concentration and the annealing temperature as already discussed. The performance of the assay is usually tested at a range of primer concentrations, from 50–900 nM using each primer at each concentration. The combination of concentrations yielding the lowest Ct is chosen (Figure 7.8).

When fluorescent oligonucleotide probes hybridise to target sequences during PCR, they must compete with the product' homologous strand. Unequal amounts of each primer may be used in an assay to generate effectively single-stranded targets, which can enable more efficient probe hybridisation since the concentration of the competing homologous PCR strand is significantly reduced. The primer that generates the homologous PCR strand is used at a significantly reduced concentration, where ratios between the two primers are typically between 10:1 and 100:1.[44] The quality of melt curve peaks may also be improved by the reduction in the amount of competing homologous product strands, although as a reduced amount of product is generated the efficiency of target detection may be reduced in some assays.

Using the optimal primer concentrations the probe/reporter is then optimised for each assay (Figure 7.9). The concentration of fluorescent probe affects the signal-to-noise ratio, so should be optimised such that the signal emitted from unhybridised probe is low or negligible, but the signal emitted from hybridised probe is significant. If too little probe is used in PCR assays, the amount of background signal is small but the amount of fluorescent signal produced upon hybridisation is also small. However, if too much probe is employed, the background fluorescence emitted from unhybridised probe will be large and may obscure the signal generated by hybridisation. The probe

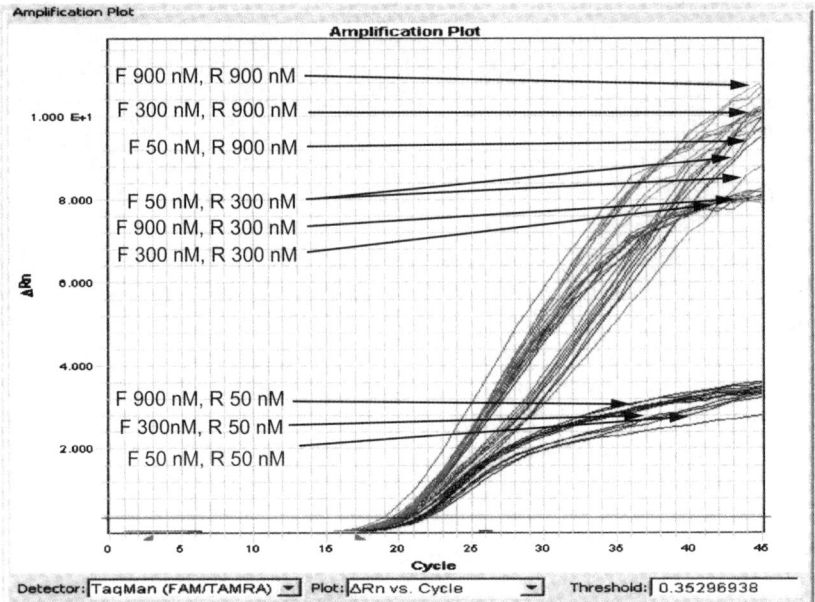

Figure 7.8 Results of a typical experiment optimising the concentration of PCR primers for a TaqMan™ assay. Manufacturer guidelines recommend that a variety of forward (F) and reverse (R) primer concentrations, usually from 50–900 nM as shown here, are tested in combination, to determine the optimal concentration for the assay. In this experiment, the 300 nM forward and 300 nM reverse combination was chosen as the optimal, being the lowest concentrations that reproducibly yielded the earliest Ct values whilst retaining a sigmoidal curve. It can also be seen that the change in reverse primer concentration has more effect on the kinetics than changes in the forward primer.

concentration also affects the hybridisation kinetics; to ensure that probe hybridisation occurs with a high efficiency and that there is sufficient probe to bind and detect the large amounts of target generated during amplification a molar excess of probe should be included in PCR reactions. However, the concentration of probe should not be sufficient to generate large background signals or to cause inhibition of the PCR.

For assays utilising fluorescent intercalating dyes as reporters, the dye concentration similarly requires optimisation for each set of PCR primers. Typically SYBR® Green I is used at a 1:10 000 dilution of the stock concentration, although higher concentrations (1:7000) may inhibit the enzyme.[45]

7.4.5 Target Level

In homogeneous assays, utilising fluorescence emission to monitor product accumulation, insufficient template may not generate increases in fluorescent

Quantitative Real-time PCR Analysis

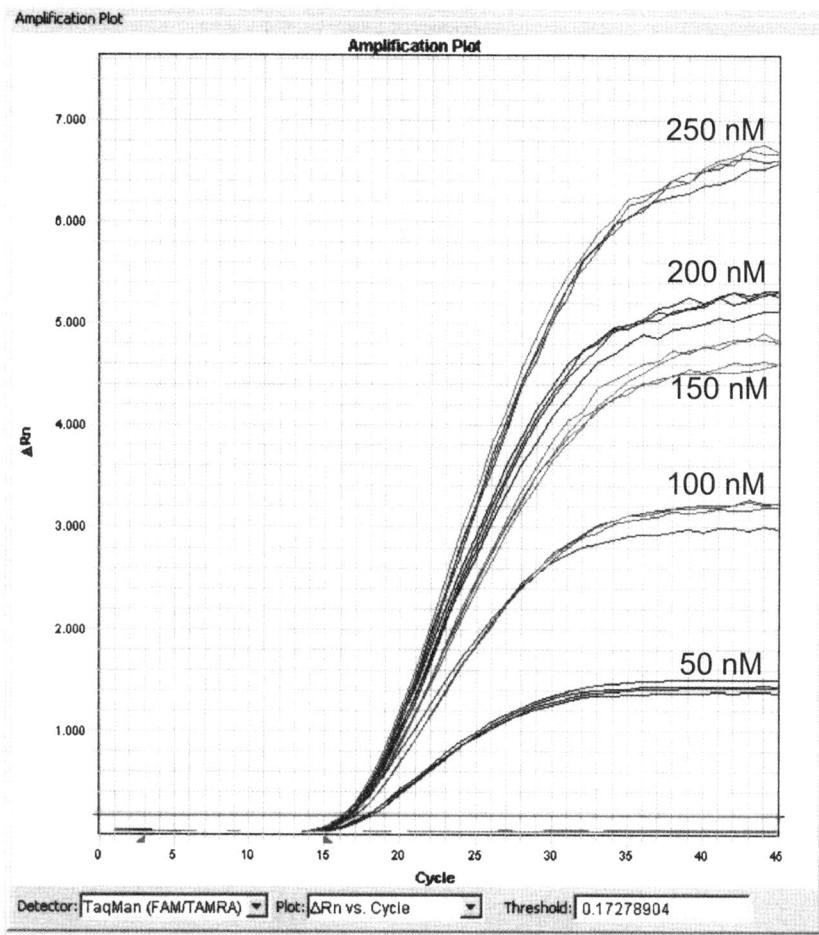

Figure 7.9 Experiment showing optimisation of a TaqMan™ probe for a typical assay. Manufacturer guidelines recommend testing a range of probe concentrations with the optimal primer levels, to determine the minimum effective amount of probe required in each assay. Here concentrations from 50 to 250 nM were tested. The 250 nM concentration was identified as yielding the highest relative fluorescence intensity, so was chosen for future use to ensure optimal assay performance.

signal during the course of amplification, whilst excess target may promote fluorescence increases prematurely in the reaction (Figure 7.10), and may also promote the generation of non-specific products. If fluorescence signal from reactions with very high template concentrations crosses the threshold value in the early cycles the baseline settings will be affected, although this can be rectified by manually setting the baseline to be calculated before any signal is detected. Typically, the quantity of genomic DNA included in homogeneous

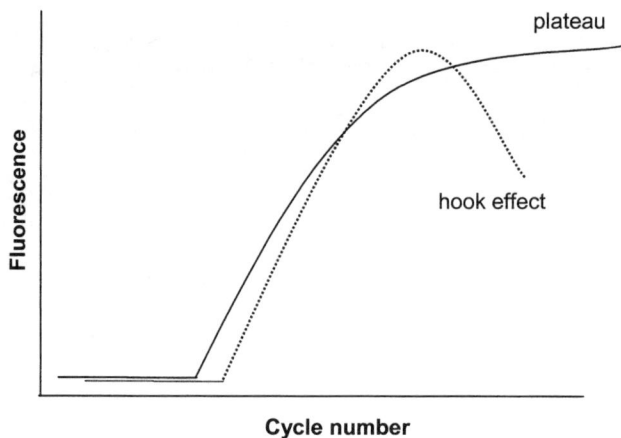

Figure 7.10 Schematic diagram illustrating the 'hook effect' often observed at high template concentrations.

assays is between 10 ng and 200 ng, although the concentration of unknown samples in quantification assays cannot always be controlled. Special considerations required when determining the amount of trace levels of target are discussed in more detail in Section 7.5.

In some experiments, a decrease in fluorescent signal is observed following the exponential phase of PCR. This 'hook effect'[46] is presumed to derive from competitive hybridisation between the single strands of the PCR product and the oligonucleotide probe. At low product concentrations, oligonucleotide probes compete efficiently for hybridisation target sites and, therefore, fluoresce efficiently. However, when the amount of PCR product is high, the two PCR strands re-anneal faster than the oligonucleotide probes can hybridise to their target sequence, such that the amount of fluorescence emission decreases. The observed decrease in fluorescence does not affect the efficiency or specificity of amplification or target detection, although optimising DNA template and $MgCl_2$ concentrations and reducing the number of PCR cycles performed, or utilising asymmetric target amplification,[47] may reduce the effect.

7.4.6 Contamination Control

As with standard PCR (Chapter 6), care is needed to ensure amplification reactions are not contaminated with exogenous targets such as amplification products from previous reactions. Homogeneous PCR assays generate significantly reduced amounts of post-amplification contaminant compared with conventional methods, since post-PCR product manipulation is not required. However, precautions such as the use of dedicated DNA-free PCR set-up areas and equipment, wearing appropriate protective clothing and using aerosol-resistant pipette tips are all still recommended. In addition, many commercial real-time master mixes include dUTP in the nucleotide mix and uracil

N-glycosylase (at approximately $0.01\,U\,\mu l^{-1}$) to clear any dUTP-containing PCR products carried over into the reaction.

7.4.7 Experimental Design

As described in Chapter 2, several factors should be considered in designing an experiment, including the appropriate level of replication, the controls to include and the need for randomisation of the samples, standards and controls. Depending on the instrument used the number of reactions that can be performed in one run can be limited (the LightCycler® 2.0, for example, has a 32 reaction capacity), and thus the appropriate choice of samples, controls and calibrants is vital to be able to interpret experimental results confidently.

7.4.7.1 Use of Controls

Negative controls should always be included, ideally both PCR reaction set-up controls and also negatives that have been subjected to the same extraction and preparation processes as the samples being analysed. It is advisable to run a number of negative controls and to intersperse their preparation with that of the unknown samples to obtain a representative estimation of the level of contamination in the analytical process. Without the inclusion of such controls it is impossible to determine if signals arise from the amplification of endogenous sample targets, or if cross-contamination between samples has occurred.

The use of positive controls is also desirable, to ensure the reaction components are functional and that the efficiency of the assay is acceptable. Internal positive control reactions can also demonstrate that no reaction inhibition has occurred,[48] which is especially important in the interpretation of apparently negative results from clinical or environmental samples. Inclusion of characterised positive samples can also be used to compare with unknown samples in post-amplification melt curve analysis, and are useful in assessing reaction specificity in genotyping assays.

7.4.7.2 Level of Replication

As discussed in Chapter 2, performing sufficient replicate measurements of an unknown sample can increase the confidence with which the quantitative data are interpreted, and is often crucial in providing sufficient analytical sensitivity for trace level analytes (Section 7.5). For quantitative determinations, six replicates are recommended to obtain a result with a low associated coefficient of variance. However, constraints of sample availability, time or cost may necessitate some reduction from this ideal. Studies of current practice in several sectors from 2004–2006 revealed that the laboratories questioned included between four and nine points in standard curves, and performed an average of three replicate reactions for both DNA standards (range 2–3) and unknown samples (range 2–8).

7.4.7.3 Randomisation

Again depending on the instrument used, it can be desirable to randomise the arrangement of samples to avoid any amplification bias resulting from temperature differences between the reaction positions. Instruments such as the Corbett Rotor-Gene™ rotate the samples during the reaction, so ensuring thermal homogeneity of the reactions. However, we have noted that the thermal cycling block on the ABI PRISM® 7700 SDS has a temperature differential, which can affect the efficiency of the reaction under certain circumstances (Figure 7.11).

The advent of higher-throughput robotic systems is facilitating the use of more complex experimental designs, such as the use of randomised plate layouts.

7.4.8 Data Analysis

Real-time PCR instruments are generally provided with manufacturer specific software, and thus the method of deriving the concentration or copy number of unknown samples from the fluorescent amplification signals may vary depending on the instrument and software version used. In addition, a number

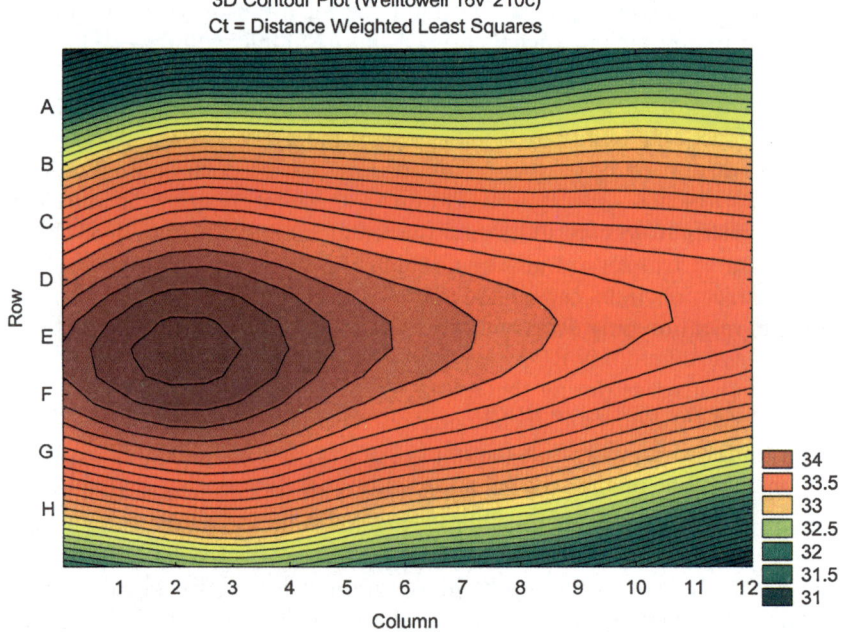

Figure 7.11 Contour plot representation of the results of 96 identical reactions set up on an 8 × 12 well microtitre plate, run on the ABI PRISM® 7700. The key shows the Ct values observed across the plate. Sub-optimal probe concentrations were used to enable differences in reaction efficiency due to temperature variations to be determined.

of quantification approaches have been detailed in the literature. It is not surprising that difficulties in data comparison may arise, despite having the same mathematical fundamentals.[49]

7.4.8.1 Basic Mathematics of PCR Amplification

The PCR process generates anywhere between an average of 0 to 1 copy of each target in each reaction cycle, so that for any cycle the number of molecules is:

$$N_C = N_{C0} \times (E + 1)^C \tag{7.1}$$

where N_C is the number of molecules at cycle N, N_{C0} is the number of molecules at cycle 0, E is the efficiency of the reaction and C is the cycle number. Making assumptions that the efficiency of the amplification is constant in the early exponential stages of the reaction, and that all standards possess the same number of target molecules at the point at which their signal crosses the determined threshold level, then the equation may be simplified to allow determination of the reaction efficiency, E_S (Equation (7.2)).

$$E_S = 10^{-\text{slope}} - 1 \tag{7.2}$$

The number of molecules at the threshold point, N_t, can also be determined from the standard curve (Equation (7.3)).

$$N_t = 10^{\text{Intercept}} \tag{7.3}$$

7.4.8.2 Data Normalisation

In many instruments each individual reaction position is excited and detected independently. In addition, in any instrument there is the potential for variability between reactions caused by small differences in pipetting. To overcome potential variation between reactions, it is possible to incorporate a passive reference dye, commonly ROX, in each reaction. The signal is monitored from the passive reference during the course of the amplification, and each well is then normalised using the unchanging reference signal. An advantage of the normalisation strategy is that the impact of any optical variability across the instrument will be minimised, but as an additional fluorescence channel is used to make the measurements the multiplexing capacity is reduced.

The raw data produced during amplification are also usually corrected to remove background noise from the measurements. It is usually possible to either set a manual baseline or to use settings calculated by the instrument. Ideally assays should be designed with a low initial signal and a significant increase resulting from amplification, to enable effective noise reduction.

A more complex normalisation approach, including curve smoothing and amplitude normalisation, has also been described to facilitate gene expression determination by a standard curve approach.[50]

7.4.8.3 Routes to Determining Amplification Efficiency

The threshold method is most commonly used for the quantification of unknowns, and as illustrated in Figure 7.1 utilises information from the points at which known DNA standards reach a specified fluorescence threshold to construct a standard curve of crossing threshold against target level. As described, the reaction efficiency is determined from the slope of the standard curve. The threshold level should be set in the exponential phase of the amplification, and most instruments calculate an optimal level setting. Manually setting the level is possible, but is subjective and may also introduce variability between runs of the assay. Advantages are that the method is simple, and the quality of the assay may be monitored using the parameters of the standard curve. Disadvantages are that the dilution series used to construct the standard curve is prone to errors, and the assumption that the reaction efficiency is a constant in the exponential phase of the reaction is not always valid.[51]

To overcome the problems in using a dilution series, alternative methods based on estimating the amplification efficiency from single reactions have been developed.[52–54] The rate of change of fluorescent signal within a single reaction may be monitored, ideally within the linear phase of signal increase, to determine the efficiency of each reaction. The second derivative maximum option in the LightCycler® software similarly calculates the maximum rate of change of the signal in the reaction, and utilises the peak to determine the fluorescence at the maxima, and hence the initial number of copies in the reaction.

A third mathematical approach utilising branching process theory to model amplification and determine reaction efficiencies has also been developed and validated using qPCR data,[55–56] and reflects the stochastic nature of the process.

7.4.8.4 Outlier Identification

Other mathematical treatment of data that can benefit the accuracy of the results obtained is identification of outliers. In ISO 5725 guidelines, outliers are classed as results which lie beyond 99% of the range of the characterised distribution (those which have a probability value less than 1%). Inclusion of such inconsistent data points can affect quantitative results, and ideally should be identified and omitted from the analysis. A number of routes to outlier identification have been developed, including detecting dissimilarities of amplification efficiencies of replicate reactions[53,57] and the use of the Grubbs test to assess Ct values.[58]

7.4.9 Validation

It may be necessary to validate a newly developed or introduced qPCR assay, to determine the scope and performance of the method in-house. The approach to method validation is described in detail in Chapter 3 and so will not be

Table 7.4 Parameters and approaches to consider in assessing qPCR performance.

Performance Characteristic	qPCR performance	Experimental procedure
Dynamic range	Range of sample concentrations over which the assay remains linear	Dilution series from a known concentration DNA analysed
Repeatability	Variability of result under closely controlled conditions	Same measurements on same sample repeated by same analyst
Reproducibility	Variability of result under differing conditions	Same sample measurement repeated by different analysts using different instruments, different laboratories or over time
Bias	Consistent over- or underestimation of the true result	Average measured value of a reference material compared to the assigned value
Specificity	Ability of the assay to detect the target but not other potential analytes present in the sample	Assay performed with a variety of related targets to check for false positive signals
Sensitivity (LOD/LOQ)	The lowest amount of the target that is detectable/reliably quantifiable	Assay performed with increasing dilutions of analyte to determine the limit of detection/linearity

repeated here, other than to highlight the performance characteristics that might be usefully assessed (Table 7.4).

The expected range for both correlation coefficient and slope of the standard curve for quantitative assays can also be determined. For TaqMan™ probe assays, with typical reaction efficiency values between 92 and 110%, the slope range would be from approximately −3.52 to −3.1. However, acceptable correlation coefficient and slope ranges may be set by the laboratory according to the needs of the application. These values can then be compared to the actual assay performance over time as measures of acceptable quality, enabling problems or errors to be identified.

7.5 Quantification of Low Levels of Target Analyte

Accurate quantification of trace amounts of DNA targets using qPCR is increasingly important for clinical, environmental and forensic applications, although instrument manufacturers generally do not recommend quantification of less than 5000 target copies.

7.5.1 Level of Variability

At high target concentrations the CV is usually less than 1%, however at low target levels the accuracy of measurement is sensitive both to target losses during sample preparation and to high levels of sampling variability. A number of studies[41,59] have shown that analytical variability increases with decreasing copy number (Figure 7.12).

7.5.1.1 Sample Handling

Solutions containing very low concentrations of target can be significantly affected by DNA sticking to tube walls during preparation stages, probably as a larger proportion of the total is lost. Sample loss may be minimised by using low retention or siliconised plastic ware, and preparing low concentration standards just prior to reaction set-up rather than storing dilute solutions.[41,61]

It is also recommended to use volumes of $\geq 250\,\mu l$ when preparing standard curve DNA dilutions for analysis, as using smaller volumes can decrease both sensitivity and precision.[62] The difference in sensitivity may be attributable either to sampling variation or to the greater surface area of the dilution solution that is in contact with the tube in the low volume dilutions. Despite the use of siliconised plasticware some such losses may still be expected, and the

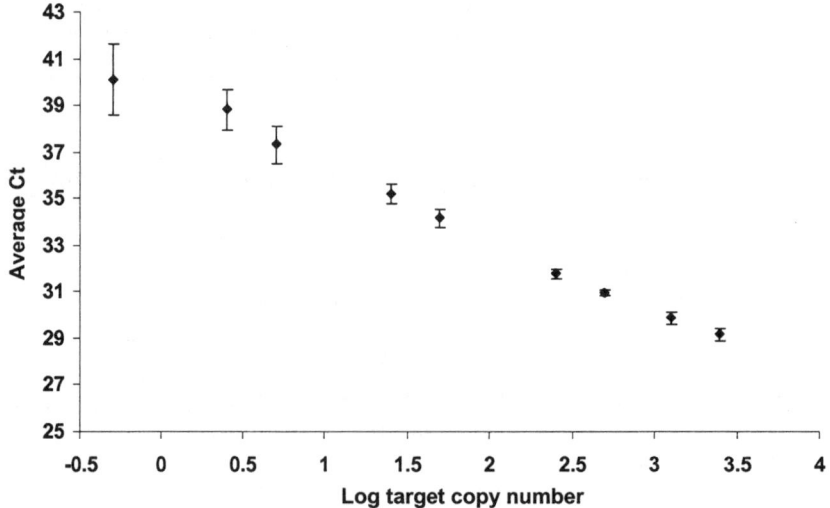

Figure 7.12 Increased variability at low target levels. The average Ct (±1 SD) values are plotted against log target copy number. The results were obtained from 12 repeat standard curves from an SRY assay[60] with all points measured in triplicate, and reactions yielding a Ct ≥ 55 excluded.

larger contact area may lead to a greater proportion of the target copies being lost through adsorption.

7.5.1.2 Amplification Cycles

Typically in the region of 30 cycles are performed in a qPCR analysis, yielding a clear measurable signal. However, assays designed to detect and quantify even very small numbers of targets can benefit from extending the number of cycles performed. Utilising a longer reaction with 55–60 cycles is of benefit both by permitting detection of samples that only reach detectable fluorescence levels late in the reaction and by maximising the difference between true negative and late-appearing positive signals (Figure 7.13). Clear distinction between reaction negative controls and positive signals can significantly increase confidence in late-appearing signals, thus increasing the effective sensitivity of the analysis.

7.5.1.3 Replication Level

Increasing the number of replicates performed can improve the effective sensitivity of an assay detecting very low concentration analyte, by raising the probability that the target will be sampled from the bulk solution. This increased sensitivity has been utilised in the clinical setting,[63] and can also be modelled using logistic regression (Figure 7.14) to determine the expected probability of target detection of an assay for a given analyte concentration.[61]

7.5.1.4 Data Handling

At very low target levels the significant proportion of replicate reactions that fail to yield a measurable Ct present an additional challenge to achieving reliable analysis. Sampling variability results in a number of low copy number reactions without any target, which consequently do not yield a signal. Such reactions preclude the use of normal statistical analysis as there is no meaningful Ct value that can be assigned. In addition, simply ignoring the

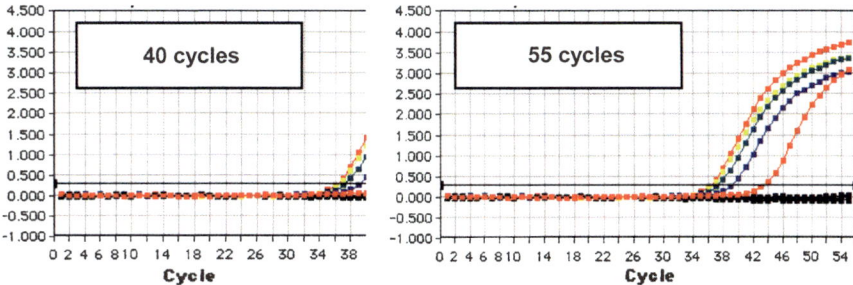

Figure 7.13 Amplification plots of SRY targets from human male genomic DNA on the ABI PRISM® 7700. Data from 40 and 55 cycle qPCR reactions, demonstrating the increased level of information obtained from very low target concentration samples (0–15 genome equivalents per reaction).

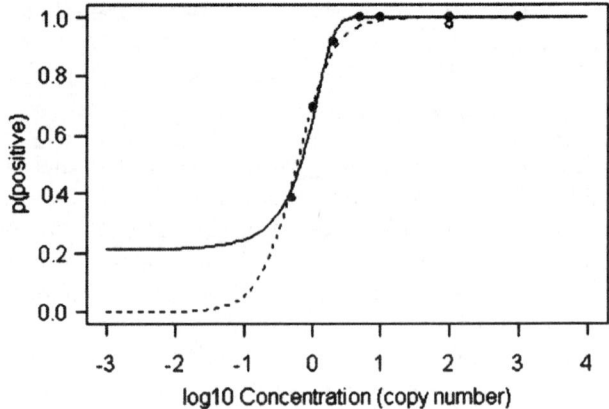

Figure 7.14 Logistic regression of detection probability.[61] Logistic regression of p(positive) vs. concentration C with false negative at $\log_{10}(C)$ omitted (solid line) and p(positive) vs. $\log_{10}(C)$ with false negative included (dashed line). Solid points show fraction of positives at each concentration with the false negative omitted; the open circle shows the calculated fraction at $\log_{10}(C)$ with the false negative included. The apparent lower limit to the solid curve is an artefact of plotting on the $\log_{10}$ axis.

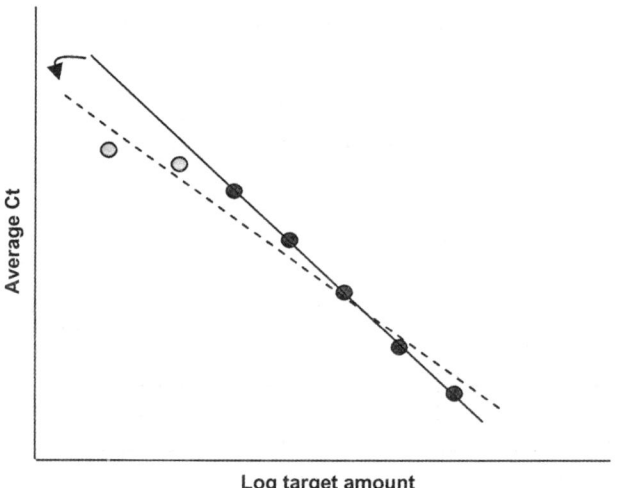

Figure 7.15 Schematic demonstrating the potential bias from omitting negative replicates and underestimating low standard concentration Ct values.[61]

replicate amplification failures is effectively omitting information from the assay, which can introduce bias into the results by assigning a lower average Ct to standards or samples than is warranted by the full data.[61] The lower Ct values can skew the regression line of the standard curve, and affect the quantification results at all target concentrations (Figure 7.15). To avoid bias in the analysis the standard curve should be constructed only using standard dilution points where all replicate reactions yield a meaningful measurable Ct value. In addition, to determine the concentration value of unknown samples average Ct values should not be used. Each replicate Ct should be separately interpolated to yield a concentration or copy number value, and any replicate reactions with no detectable signal due to sampling variability at very low concentration can be assigned a value of zero. Then the concentration of target in the unknown sample may be determined by averaging the replicate reaction values.

7.6 Standards and Comparability

The development of qPCR has enabled performance of quantitative measurements with very low associated variability, with CVs of between 2 and 5% commonly reported in the literature. However, there are many potential sources of analytical error in the process, ranging from operator-introduced variability,[64] temperature or detector heterogeneity in the instrument, variations in reagent integrity and activity, effects of target melting behaviour[65] and inherent variations in the PCR amplification itself. The combination of these factors can have a significant cumulative effect on quantitative results obtained, which may be further exacerbated by the exponential nature of the reaction. Consequently it is desirable for users of the technique to understand the sources of experimental variability, in order to control and/or monitor for potential problems.

7.6.1 Quantitative Standards

The accuracy of any qPCR analyses that utilise DNA standards or calibrators to anchor the quantitative measurements is largely dependent on the quality of the standard used. Commonly DNA standards are prepared and quantified in-house, largely because suitable reference materials are not yet commercially available. The quantification of DNA standards may vary significantly between laboratories and methods, as discussed in more detail in Chapter 5. Thus the ability to perform 'absolute quantification' using real-time PCR analysis is practically limited by the certainty with which the concentration of the DNA standard is known. However, the software provided with many real-time instruments generates concentration or copy number results to several decimal places, implying a degree of analytical accuracy that is unlikely to be achieved in practice.

Of perhaps more concern is that the approximate DNA concentrations assigned to many commercial DNA preparations are sometimes used as accurate concentration values by researchers. The downstream analyses are thus often reliant on concentration values which are neither intended nor suitable for the purpose. In the absence of certified quantitative reference materials and standards, an awareness of the limitations of quantitative accuracy is important to ensure results are understood and interpreted correctly.

Assays that depend on relative quantification are unaffected by the lack of quantitative DNA standards, as the ratio of two targets within the assay is assessed. If both targets are detected within the same assay, then many of the sources of measurement uncertainty similarly affect both determinations, and thus do not influence the final result. As an example, assays designed for the relative quantification of genetically modified organism (GMO) utilise the ratio of Ct values of endogenous and GM sequences. For this application reference materials are available containing a certified percentage GM, which are typically used as comparative standards underpinning confidence in the assay.

7.6.1.1 Instrument Calibration

Variability may be introduced by non-uniformity within the thermal cycler itself, depending on the instrument design. Consequently it is recommended to check the instrument performance regularly, to ensure the machine has an established calibration schedule and that any maintenance recommended by the manufacturer is carried out. Several commercial dye calibration kits and systems are currently available, which are useful to ensure that the selectivity of signal detection is set correctly. In addition, checks to ensure that the well or rotor positions are not contaminated with extraneous fluorescent dye and that the instrument is detecting fluorescence data at the correct positions are possible on some machines.

In addition it is possible for the thermal uniformity of heating blocks to be assessed by accredited testing laboratories, and a system for interim checks is also now available (the DRIFTCON® system from Anachem).

7.6.1.2 Comparability

Although qPCR exhibits low levels of variability when performed within a laboratory, comparative performance between laboratories is also required for many applications. Comparable performance may be achieved by using standardised methods and data handling approaches, calibrated equipment and reference materials where available. Several inter-laboratory performance studies have been performed utilising real-time PCR, and the results have demonstrated that there is appreciable variability between results obtained in different laboratories.[66,67] An early study of 42 clinical diagnostic laboratories found variation in quantitative accuracy, assay precision and sensitivity. In a later study of over 130 laboratories using a TaqMan™ based assay, requiring RNA extraction, reverse-transcription and qPCR, about one in five laboratories submitted one or more unsatisfactory results.[66]

Figure 7.16 shows the results of a quantitative TaqMan™-based assay performed by independent analysts as part of a proficiency testing scheme. The reported results varied by over an order of magnitude, highlighting the potential variability in quantitative results generated by different laboratories and analysts.

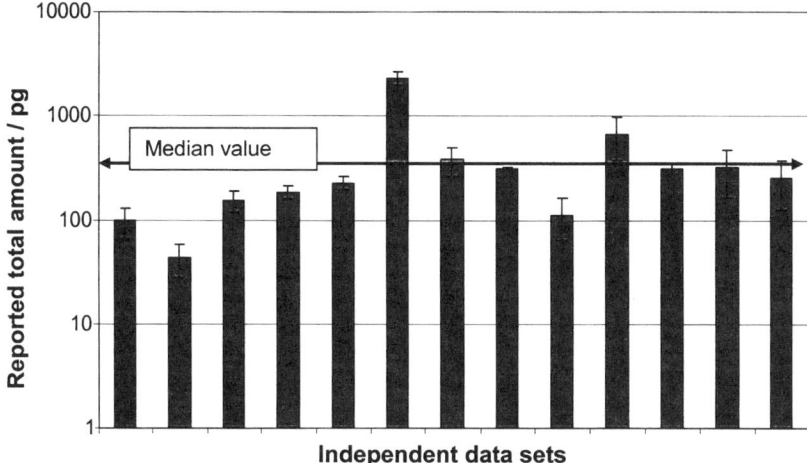

Figure 7.16 Results from one round of a PT scheme using a TaqMan™-based qPCR assay. The data have been normalised to exclude variability arising from DNA standard measurements, and reflect the average values of three identical samples provided to participants. The error bars represent ±1 SD of the three results for each participant, and the median value is shown as a solid line.

7.6.1.3 Measurement Uncertainty

Increasingly there is a requirement for analytical data to be reported with an associated measurement uncertainty value, as introduced in ISO 17025:2005. Detailed information on the calculation of uncertainty for a method can be found in the ISO Guide to the Expression of Uncertainty in Measurement (GUM). The need for determination of the uncertainty associated with a quantitative measurement arises because all measurements are estimations in reality, because it is not possible to completely control all possible sources of variability in an analysis to obtain a 'true value'. In summary the variability for each individual stage of the qPCR process should be determined, documented and combined to produce an overall uncertainty budget. In practice this is difficult to achieve, and estimation may be made from validation data and inter-laboratory comparisons.

7.7 Summary

As PCR has developed from a research to an analytical tool there has been an increasing demand for the provision of accurate quantitative data. Despite variation in the details of the approaches and assays that have been developed, certain principles underpin all such analyses in striving to achieve analytical accuracy. The use of appropriate controls and standards is essential in the quantitative process and should be selected to control for sample type and history, and be equivalent in terms of amplification potential. Careful assay design in order to ensure equivalence of analytes and standards, and care in performing each assay to minimise potential variability resulting from pipetting, data analysis and documentation, will additionally serve to improve precision. Whilst technological improvements have been made allowing quantitative PCR measurements to be approached by the analytical community, the persisting challenge to the analyst is to demonstrate the accuracy of such measurements. Provision of suitable standards, certified reference materials and other QA tools such as appropriate accessible proficiency trials may help overcome current problems, and allow qPCR to fulfil its full analytical potential.

References

1. G. Gilliland, S. Perrin, K. Blanchard and H. F. Bunn, *Proc. Natl. Acad. Sci. USA*, 1990, **87**, 2725.
2. D. A. Levinson, J. Campos-Torres and P. Leder, *J. Exp. Med.*, 1993, **178**, 317.
3. M. J. Alvarez, A. M. Depino, O. L. Podhajcer and F. J. Pitossi, *Anal. Biochem.*, 2000, **287**, 87.
4. R. Higuchi, G. Dollinger, P. S. Walsh and R. Griffith, *Biotechnology (NY)*, 1992, **10**, 413.
5. http://www.stratagene.com/lit_items/TN5_Standard_Curves_technotefinal.pdf (accessed May 2007).
6. L. G. Lee, C. R. Connell and W. Bloch, *Nucl. Acids Res.*, 1993, **21**, 3761.
7. P. M. Holland, R. D. Abramson, R. Watson and D. H. Gelfand, *Proc. Natl. Acad. Sci. USA*, 1991, **88**, 7276.
8. K. J. Livak, S. J. Flood, J. Marmaro, W. Giusti and K. Deetz, *PCR Meth. Appl.*, 1995, **4**, 357.
9. K. J. Livak, *Genet. Anal.*, 1999, **14**, 143.
10. http://www.appliedbiosystems.com/ (accessed May 2007).
11. I. V. Kutyavin, I. A. Afonina, A. Mills, V. V. Gorn, E. A. Lukhtanov, E. S. Belousov, M. J. Singer, D. K. Walburger, S. G. Lokhov, A. A. Gall, R. Dempcy, M. W. Reed, R. B. Meyer and J. Hedgpeth, *Nucl. Acids Res.*, 2000, **28**, 655.
12. C. Letertre, S. Perelle, F. Dilasser, K. Arar and P. Fach, *Mol. Cell. Probes.*, 2003, **17**, 307.
13. S. Tyagi and F. R. Kramer, *Nat. Biotechnol.*, 1996, **14**, 303.
14. http://www.molecular-beacons.com/default.htm (accessed May 2007).
15. S. Tyagi, D. P. Bratu and F. R. Kramer, *Nat. Biotechnol.*, 1998, **16**, 49.
16. D. J. French, C. L. Archard, T. Brown and D. G. McDowell, *Mol. Cell. Probes.*, 2001, **15**, 363.
17. M. Lee, S. Siddle and R. Hunter, *Anal. Chim. Acta*, 2002, **457**, 61.
18. C. T. Wittwer, M. G. Herrmann, A. A. Moss and R. P. Rasmussen, *Biotechniques*, 1997, **22**, 130.
19. P. S. Bernard, R. S. Ajioka, J. P. Kushner and C. T. Wittwer, *Am. J. Pathol.*, 1998, **153**, 1055.

20. D. Whitcombe, J. Theaker, S. P. Guy, T. Brown and S. Little, *Nat. Biotechnol.*, 1999, **17**, 804.
21. S. C. Johnson, C. B. Sherrill, D. J. Marshall, M. J. Moser and J. R. Prudent, *Nucl. Acids Res.*, 2004, **32**, 1937.
22. H. A. Cubie, A. L. Seagar, E. McGoogan, J. Whitehead, A. Brass, M. J. Arends and M. W. Whitley, *Mol. Pathol.*, 2001, **54**, 24.
23. S. D. Dufresne, D. R. Belloni, W. A. Wells and G. J. Tsongalis, *Arch. Pathol. Lab. Med.*, 2006, **130**, 185.
24. http://bio-rad.com (accessed May 2007).
25. http://www.stratagene.com (accessed May 2007).
26. http://www.techne.com (accessed May 2007).
27. http://www.biogene.com (accessed May 2007).
28. http://biochem.roche.com (accessed May 2007).
29. http://www.cepheid.com/Sites/cepheid/section.cfm?id=1 (accessed May 2007).
30. http://www.corbettlifescience.com (accessed May 2007).
31. L. J. Kricka and P. Wilding, *Anal. Bioanal. Chem.*, 2003, **377**, 820.
32. http://web.ncifcrf.gov/rtp/gel/primerdb/ (accessed May 2007).
33. M. Zuker, *Nucl. Acids Res.*, 2003, **31**, 3406.
34. http://www.bioinfo.rpi.edu/applications/mfold/ (accessed May 2007).
35. http://www.ncbi.nlm.nih.gov/BLAST/ (accessed May 2007).
36. http://www.biocenter.helsinki.fi/bi/Programs/fastpcr.htm (accessed May 2007).
37. http://www.autoprime.de/AutoPrimeWeb (accessed May 2007).
38. R. B. Wallace, M. J. Johnson, T. Hirose, T. Miyake, E. H. Kawashima and K. Itakura, *Nucl. Acids Res.*, 1981, **9**, 879.
39. Z. Guo, Q. Liu and L. M. Smith, *Nat. Biotechnol.*, 1997, **15**, 331.
40. J. G. Wetmur, *Annu. Rev. Biophys. Bioeng.*, 1976, **5**, 337.
41. I. A. Teo, J. W. Choi, J. Morlese, G. Taylor and S. Shaunak, *J. Immunol. Meth.*, 2002, **270**, 119.
42. K. A. Kreuzer, A. Bohn, U. Lass, U. R. Peters and C. A. Schmidt, *Mol. Cell. Probes*, 2000, **14**, 57.
43. C. Hilscher, W. Vahrson and D. P. Dittmer, *Nucl. Acids Res.*, 2005, **33**, e182.
44. S. K. Poddar, *Mol. Cell. Probes*, 2000, **14**, 25.
45. K. Nath, J. W. Sarosy, J. Hahn and C. J. Di Como, *J. Biochem. Biophys. Meth.*, 2000, **42**, 15.
46. Roche Molecular Biochemicals Technical note, No. LC 8/99, 1999.
47. K. Barratt and J. F. Mackay, *J. Clin. Microbiol.*, 2002, **40**, 1571.
48. M. Stöcher, V. Leb, G. Holzl and J. Berg, *J. Clin. Virol.*, 2002, **25**, Suppl 3, S47.
49. R. G. Rutledge and C. Cote, *Nucl. Acids Res.*, 2003, **31**, e93.
50. A. Larionov, A. Krause and W. Miller, *BMC Bioinformatics*, 2005, **6**, 62.
51. W. Liu and D. A. Saint, *Anal. Biochem.*, 2002, **302**, 52.
52. W. Liu and D. A. Saint, *Biochem. Biophys. Res. Commun.*, 2002, **294**, 347.

53. S. N. Peirson, J. N. Butler and R. G. Foster, *Nucl. Acids Res.*, 2003, **31**, e73.
54. A. Tichopad, M. Dilger, G. Schwarz and M. W. Pfaffl, *Nucl. Acids Res.*, 2003, **31**, e122.
55. J. Peccoud and C. Jacob, *Biophys. J.*, 1996, **71**, 101.
56. N. Lalam, C. Jacob and P. Jagers, *Adv. Appl. Probab.*, 2004, **36**, 602.
57. T. Bar, A. Stahlberg, A. Muszta and M. Kubista, *Nucl. Acids Res.*, 2003, **31**, e105.
58. M. J. Burns, G. J. Nixon, C. A. Foy and N. Harris, *BMC Biotechnol.*, 2005, **5**, 31.
59. J. Stenman and A. Orpana, *Nat. Biotechnol.*, 2001, **19**, 1011.
60. L. Birch, C. A. English, K. O'Donoghue, O. Barigye, N. M. Fisk and J. T. Keer, *Clin. Chem. (Washington, DC)*, 2005, **51**, 312.
61. S. L. Ellison, C. A. English, M. J. Burns and J. T. Keer, *BMC Biotechnol.*, 2006, **6**, 33.
62. C. English, L. Birch and J. Keer, LGC Report, 2003, **LGC/VAM/2003/057**, available at www.nmschembio.org.uk through free registration.
63. I. Hromadnikova, B. Houbova, D. Hridelova, S. Voslarova, J. Kofer, V. Komrska and D. Habart, *Prenat. Diagn.*, 2003, **23**, 235.
64. S. A. Bustin, *J. Mol. Endocrinol.*, 2002, **29**, 23.
65. J. Wilhelm, M. Hahn and A. Pingoud, *Clin. Chem. (Washington, DC)*, 2000, **46**, 1738.
66. S. C. Ramsden, S. Daly, W. J. Geilenkeuser, G. Duncan, F. Hermitte, E. Marubini, M. Neumaier, C. Orlando, V. Palicka, A. Paradiso, M. Pazzagli, S. Pizzamiglio and P. Verderio, *Clin. Chem. (Washington, DC)*, 2006, **52**, 1584.
67. C. C. Raggi, P. Verderio, M. Pazzagli, E. Marubini, L. Simi, P. Pinzani, A. Paradiso and C. Orlando, *Clin. Chem. Lab. Med.*, 2005, **43**, 542.

CHAPTER 8
Multiplex PCR and Whole Genome Amplification

LYNDSEY BIRCH, CHARLOTTE L. BAILEY AND MORTEN T. ANDERSON

LGC, Queens Road, Teddington, TW11 0LY

In this chapter two versatile techniques are described, which are utilised for simultaneous amplification of multiple DNA targets. Multiplex PCR (mPCR) is a method that simultaneously co-amplifies from two or more primer pairs from a single sample. In a further extension of multiplexed amplification, whole genome amplification (WGA) methods have been developed to increase the amount of DNA targets in a sample without bias, usually in order to facilitate downstream analyses. The basic principles of both techniques are described in the following sections, together with some considerations underpinning reliable multiplex PCR development.

8.1 Introduction to Multiplex PCR

Many experimental approaches require analysis of a variety of DNA sequences, necessitating multiple PCRs to be performed on the same or related templates (for example, comparison of different regions of the same gene, or comparison of different genomes for speciation). Considerable savings of time, effort and reagent costs can be achieved by simultaneously amplifying multiple sequences in a single reaction.[1] Whilst there is no theoretical limit to the number of loci which can be amplified simultaneously, there are a number of practical constraints to be considered, which place limits on the technique.[2]

Firstly, as the number of loci increases it becomes increasingly difficult to balance the optimal PCR conditions for all reactions. Even small variations in the efficiency of PCR amplification across the loci can lead to marked differences in product yield, and this may present difficulties in the detection and interpretation of all the targets in a multiplex. Another limitation is the increasing chance of non-specific amplification between non-paired primers.

As the number of primers in the reaction increases, the permutations of primer pairs which may interact non-specifically increases significantly. The amplification of significant amounts of non-specific product may hinder or prevent a clear interpretation of the results of some or all of the desired reactions. Finally, the detection system employed must be able to unambiguously identify the products from every locus amplified in the mPCR. As the number of loci increases, the differentiation of these products can become more difficult, irrespective of the detection system employed.

Multiplex PCR is generally divided into two groups:

- Linked mPCR, where different regions of the same gene or genome are amplified as exemplified by STR profiling for forensic analysis;
- Non-linked mPCR, where sequences located on unrelated but co-extracted or premixed genomes are amplified, for example in the identification of different bacterial species.[3]

Multiplex PCR can be an end-point analysis in itself or it can be a preliminary stage leading to applications such as forensic analysis,[4,5] array-based assays,[6] SNP detection/genotyping,[7] sequencing, hybridisation and restriction digest analysis.[8,9]

8.1.1 Number of Targets Amplified During Multiplex PCR

As may be expected, multiplex PCR optimisation becomes increasingly difficult as the number of targets to be amplified in a single reaction increases. In situations where it is necessary to amplify several hundred targets prior to hybridisation onto an array for example, it is often necessary to split large multiplexes into several smaller multiplexes.[10,11] Regardless of the number of targets to be amplified in an mPCR, it is important to optimise fully the reaction(s); this will be discussed in further detail in the sections below.

8.2 Design and Optimisation of mPCR

Design and optimisation of mPCR is extremely challenging and laborious,[12–14] however with the development of web-enabled systems such as MuPlex[7] the process has become more straightforward. With a system such as MuPlex the analyst provides a set of DNA sequences along with primer selection criteria, interaction parameters and the level of target multiplexing. The software then designs a set of multiplex PCR assays that cover as many of the input sequences as possible.

8.2.1 Design Strategy

An ideal mPCR would be one in which all primer pairs amplify their specific targets with equal efficiency. However, in practice this is fairly difficult to achieve and an acceptable compromise in amplification efficiency must be reached for all loci in the reaction. The development of an efficient multiplex requires careful planning and often requires extensive testing to determine optimal reaction conditions. Several key factors must be considered and these will be discussed in more detail below.

8.2.2 Amplification Target

The amount of template DNA that is added to a PCR can affect the sensitivity and specificity of the reaction, and may need to be optimised for each assay. False negative results may occur if insufficient DNA is used; for example when performing genotype analysis, too little template DNA may cause allelic dropout due to the preferential amplification of one variant, which may result in a heterozygous sample being reported as a homozygote. In contrast, excess template DNA may lead to non-specific amplification or even inhibition of the PCR. When performing mPCR, higher template DNA concentrations may be required to ensure that there is efficient amplification of all the target sequences within the multiplex.

The regions chosen for amplification will essentially be determined by the nature of the analysis. For example, while microbial identification assays may target species or strain specific variations or toxin genes, forensic assays will distinguish individual variations at highly polymorphic loci.

8.2.3 Primer Positioning

Several factors should be considered when deciding where to position primers for an mPCR system. Firstly, the sequence of the flanking regions of the target site may impose constraints; either that insufficient flanking sequence is available to give any flexibility in primer position, or that flanking regions contain sequences unsuitable for primer sites, such as non-unique or repetitive sequences, or sequences with high or low GC content.

Secondly, the detection of the multiplexed products must also be considered. If gel electrophoresis is to be used as the end-point detection system it is important to position primers such that the fragments may be easily separated by size, although the size range should not be so great that they cannot be resolved on a single gel. Other detection systems may not be dependent on difference in fragment size to distinguish products, allowing identical or very similar sized products to be identified. For example, the use of fluorescent-dye-based detection systems, even if electrophoresis based, may allow two or more co-migrating products to be analysed, provided they are labelled with different coloured fluorescent dyes. Other systems, such as solid phase capture methods may impose no significant restrictions on product sizes and so allow greater freedom in primer position and choice (Section 8.3).

The different efficiencies in amplification of loci within a multiplex are also an issue when positioning mPCR primers. A large size range in the expected products may well lead to preferential amplification of smaller PCR products over larger ones. In extreme cases this can result in the failure of one or more

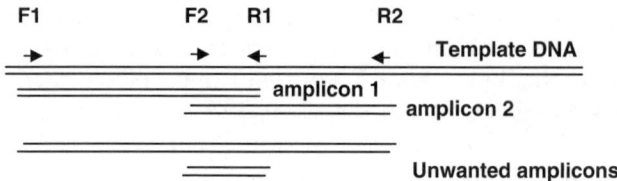

Figure 8.1 Position of priming sites in linked mPCR. The positioning of the two primer sets (F1/R1 and F2/R2) results in the desired amplicons (1 and 2) and also two undesired amplicons, primed by F1 and R2 and F2 and R1 respectively.

loci to amplify. In addition, when developing linked mPCRs it is worth noting that overlapping targets will result in a mix of nested products rather than solely the two desired amplicons, and therefore such positioning should be avoided (Figure 8.1). When high specificity is required it is important to choose priming sites from variable regions, allowing a site unique to the target of interest to be chosen. It is important to ensure that there are no published SNPs underlying the primer/probe sites, as this could cause genotypes to be miscalled and may imbalance the multiplex reaction.[15–17] For example, the universal bases inosine, 3-nitropyrrole and 5-nitroindole can be used for neutralisation of unwanted polymorphisms underlying primer sequences.

The more sequence data that are available for potential priming sites, the easier it is to develop compatible primer sets.

8.2.4 Primer Design

Primer design for multiplex PCR is governed by the same rules as for single PCR amplifications, but additional factors must be considered to ensure that a compatible primer set is obtained. An important first step in designing a multiplexed reaction is to consider the predicted melting temperatures (T_m), the temperature at which 50% of the template annealing sites and primers are in duplex. Primers should be designed to ensure that each primer within a primer set should have a similar T_m. Melting temperature values are best calculated using one of the many commercial primer design packages currently available, such as Visual OMP 6[18] and Primer Express®.[19] In general, empirical indicators are that the primers should be as uniform in both length and GC content as possible. GC distribution along the length of the fragments should also be uniform, with no high-melting temperature regions.

It should be noted that calculated T_m values should not be taken as being completely reliable indicators of primer behaviour. Rather, they should be seen as a starting point from which optimisation of performance can proceed. If one locus in a multiplex system is later seen to result in low product yield, or is identified as a source of artefact products, it may be that the true T_m is significantly different from that calculated and it will be necessary therefore to redesign one or both of the primers to overcome the problem.

Primers should also be checked for secondary structure and to ensure low homology, particularly at the 3' end, both within sets and between sets, to limit the generation of primer dimers and other artefacts, such as hairpin structures. Information of this kind can also be obtained from primer design software packages such as Primer Express®[19] and mfold.[20] An mPCR may be generated by combining published primer sets specific to the regions of interest, but designing sets specifically to work together gives greater flexibility and can avoid many problems.

8.2.5 Standardisation of Oligonucleotide T_m

8.2.5.1 Base Analogues

Occasionally, the choice of target sequence for analysis is very limited, possibly being located in very AT or GC rich regions of DNA. Therefore, the numerous oligonucleotides that are required to simultaneously analyse multiple target sequences may possess considerably different melting temperatures. To equalise the T_ms of these different sequences, various lengths of oligonucleotide may be employed. Alternatively, the melting temperatures of the numerous oligonucleotides required for multiplex and multi-measurement DNA analysis could be standardised through the incorporation of DNA base analogues.

Certain modified nucleotides have been demonstrated to stabilise and destabilise DNA duplexes when substituted for standard Watson–Crick bases.[21,22] Table 8.1 describes several of the most commonly used analogues and their effects on T_m.

8.2.5.2 Peptide Nucleic Acid (PNA)

Peptide Nucleic Acid (PNA) is an analogue of DNA in which the backbone is a pseudopeptide rather than a sugar (see Figure 8.2). An uncharged, achiral backbone is made from N-(2-aminoethyl)-glycine units linked by amide bonds. The four standard nucleobases, adenosine, cytosine, guanine and thymine, are attached to the secondary amine *via* a methylene carbonyl linkage. PNA mimics the behaviour of DNA and binds complementary nucleic acid strands. The neutral backbone of PNA results in stronger binding, due to the lack of charge repulsion and greater specificity than normally achieved because of its high thermal stability resulting in an increased T_m (approximately 1 °C higher per base pair for PNA/DNA duplex compared to DNA/DNA duplex).[31,32]

The uncharged PNA structure creates stronger binding independent of salt concentration, allowing more robust hybridisation applications. PNA oligomers also have resistance to nucleases and proteases.

8.2.5.3 Locked Nucleic Acid (LNA)

Locked Nucleic Acid (LNA) is a novel class of nucleic acid analogue that may also be used to alter the hybridisation characteristics of oligonucleotide primers and probes. LNA monomers are bicyclic compounds structurally similar to RNA nucleosides. The term 'Locked Nucleic Acid' has been coined to emphasise that the furanose ring conformation is restricted in LNA by a methylene linker that connects the 2'-O position to the 4'-C position (Figure 8.3). LNA

Table 8.1 Effect of substitution of DNA base analogues on oligonucleotide T_m.

Modified nucleotide	Substituted for	Average effect on hybridisation T_m
2-Amino-dU	dA	+3.0 °C per substitution[23]
2-Amino-dA	dU	+3.0 °C per substitution[23]
5-Methyl-dC	dC	+3.0 °C per substitution[23]
C5-Propyne-dC	dC	+2.8 °C per substitution[23]
C5-Propyne-dU	dU	+1.7 °C per substitution[23]
N4-Ethyl-dC	dC	Decreases stability of GC base-pairs, making them approximately equivalent to AT base-pairs[23]
5-Nitroindole	Any (universal base)	Does not hybridise significantly to any Watson–Crick bases causing uniform duplex destabilisation[24]
Inosine	Any (universal base)	Provides hydrogen bonds for all four Watson–Crick bases. Possibly providing T_m standardisation[25]
5-Nitroindazole	Any (universal base)	Stable hybridisation to all four Watson–Crick nucleotides with small variations in T_m between the different bases[26]
Tricyclic Aminoethyl-Phenoxazine 2′-deoxyCytidine analogue (AP-dC)-G-Clamp1	dC	+18 °C per substitution AP-dC should stabilise a duplex due to its ability to interact with both Watson–Crick and Hoogsten faces of the target dG[27]
Capped phosphoramidites-5′-Trimethoxystilbene Pyrenylmethylpyrrolindol	Attached to any base during synthesis on a solid support	+10 °C per modification. The caps favour the formation of stable Watson–Crick duplexes by stacking on the terminal base[28,29]
Locked Nucleic Acid (LNA)	dA, dT, dC, dG	+3–8 °C per substitution[30]

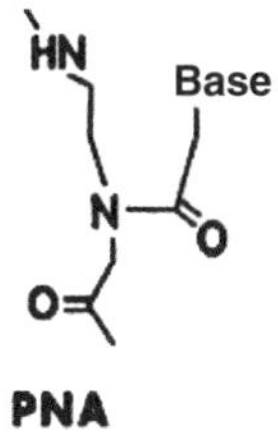

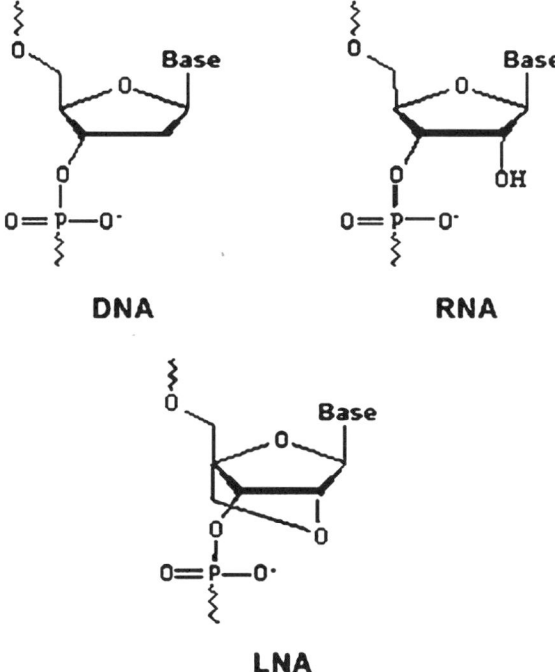

Figure 8.2 Structure of PNA. Peptide Nucleic Acid is an analogue of DNA in which the sugar backbone is replaced by a pseudopeptide.

Figure 8.3 Structure of LNA compared to DNA and RNA. In LNA, the furanose ring conformation (seen in DNA and RNA) is restricted by a methylene linker that connects the 2'-O position to the 4'-C position.

oligomers obey Watson–Crick base pairing rules and hybridise efficiently and specifically to complementary target sequences. LNA has the highest affinity towards complementary DNA and RNA ever reported. In general, the thermal stability of an LNA/DNA duplex is increased between 3 °C and 8 °C per modified base in the oligonucleotide. LNA nucleotides may be incorporated into the primers utilised in multiplex PCR.

Due to their effect on T_m, primers of shorter lengths can be used when amplifying with PNA or LNA. These modified nucleic acids may be particularly useful when amplifying related pseudogenes, as they improve amplification specificity.

The intentional inclusion of nucleotide mismatches has the practical effect of destabilising the primer-target hybrid and thus reducing the T_m. However, the specificity of the PCR may be compromised as a result, and thus alternative approaches to standardising T_ms, as described here, are recommended.

8.2.6 Optimisation

When optimising reaction components and cycling parameters, and when initially designing mPCR, an empirical approach is recommended. Time invested in understanding and optimising all the steps of the PCR cycle will certainly result in a more efficient and robust assay.

8.2.6.1 Initial Assay Development

PCR conditions should first be optimised empirically for each primer set individually to establish their sensitivity and specificity. Particular attention should be paid to optimal annealing temperature and time using standard buffering conditions. Primer sets that do not function well under standard conditions are best replaced.

Having established optimal conditions for each primer set, a single central set should be chosen and sets sequentially added to establish initially a duplex, then a triplex and so on. Clearly, the optimum conditions for each primer set may vary so it is important to continue an empirical process of optimisation as more primer sets are added. The aim is to end up with a set of conditions that represent an acceptable compromise as far as the amplification efficiencies of the component reactions are concerned. An overall decrease in sensitivity of the reaction may be observed on development of the multiplex. This may be due in part to sub-optimal amplification conditions and also, in the case of non-linked mPCR, may be due to an overall increase in the level of non-target nucleic acid and compounded impurities in the reaction.

To obtain uniform amplification of all reactions within a multiplex, variation in primer concentrations between primer sets may be required. This is particularly evident where there are variable target copy numbers but also if the reaction contains primer sets with highly variable primer/target annealing efficiencies.

Other factors which may require consideration are: buffer composition (particularly Mg^{2+} concentration); template quantity and quality; polymerase type and source; reaction volume; performance of and transferral between thermal cyclers; benefit of hot-start or touchdown PCR; number of cycles and temperature; time and ramp rates of all steps within the PCR cycle, many of which are discussed further below.

8.2.6.2 Reaction Components

Due to the compromises that are necessarily made to reaction parameters to produce simultaneous amplification of multiple sequences, the resulting mPCR may lack the robustness required for its desired application. Careful manipulation of reaction components can result in an increase in robustness, aiding the repeatability, reproducibility and sensitivity of the reaction.

Increasing magnesium ion, nucleotide and enzyme concentration may be necessary on increasing the number of targets within the reaction, considering that the total quantity of amplified product generated within an mPCR is usually greater than that generated in a uniplex reaction. In a standard PCR, 1.5 mM $MgCl_2$ is regularly utilised. However, mPCR assays require higher concentrations of $MgCl_2$, frequently 3–5 mM $MgCl_2$ is used for efficient target amplification. Generally, high dNTP concentrations should be used in mPCR to avoid reagent exhaustion; 0.25 mM dNTPs are recommended for most mPCR assays, although this may vary in each individual case. A high concentration of DNA polymerase is required in mPCR to prevent less-efficient reactions being out-competed. However, using too much DNA polymerase may also result in reduced specificity leading to the formation of spurious PCR products. A concentration of 0.05 U μl^{-1} *Taq* DNA polymerase is recommended for mPCR, although different enzyme concentrations may be required for each individual case.

Due to the requirement of the polymerase enzyme for free magnesium ions and the chelation of magnesium ions by nucleotides, an increase in nucleotide concentration without a related increase in magnesium ion concentration can rapidly inhibit the PCR. Magnesium ion concentration, in particular, should be titrated carefully as the resulting decrease in specificity on raising the concentration can often outweigh the benefits of increased amplification.

The quality and quantity of template can affect the robustness of even the most carefully designed mPCR. Due to the often sub-optimal nature of the multiplexed reaction, the presence of any inhibitory agents within the nucleic acid extract is tolerated poorly, leading to limited amplification, or indeed complete loss of the desired products. It may prove necessary to carry out an additional 'clean up' or dilution step during sample preparation when the extract is to be used in an mPCR. When using non-linked PCR to amplify template DNA isolated from different organisms, it may be necessary to investigate the relative efficiency of DNA recovery from those organisms. Ideally the DNA extraction technique will isolate DNA with equal efficiency from all target sources.

A variety of buffer additives and enhancing agents have been reported to increase PCR yield, hybridisation and amplification specificity and assay reproducibility (see Chapter 5). Whilst these additives may convey beneficial effects to particular assays, it is not yet possible to predict which agents will be useful when designing specific assays. Therefore, buffer additives must be tested empirically for each combination of template, primers and probes. PCR adjuncts, such as betaine and TMAC, which reduce or even eliminate the relationship between nucleotide composition and sequence melting temperature, may aid the development of successful multiplex assays.

8.2.6.3 Cycling Parameters

As previously stated, the annealing time and temperature of the reaction have a significant effect on assay performance and so need to be considered early in the development of the reaction. As the number of amplification products increases, there must be a corresponding increase in extension time. This ensures the full extension of targets with longer length products and the efficient amplification of primer sets with sub-optimal annealing to the target. The minimum number of cycles that allows efficient and robust detection of all fragments of interest should be used for the assay. Additional cycles can lead to difficulties in interpretation of results, especially if the reaction has been designed to be semi-quantitative.

The means by which the non-templated base addition (see Section 8.2.9) is encouraged is by the incorporation of an extended final extension hold at the end of the PCR. Classically, this has been carried out at the optimal temperature for DNA *Taq* polymerase extension (72 °C), although it has been reported that a 60 °C final extension step encourages more complete addition. Once again, an empirical approach to optimisation is recommended as different systems are likely to display different efficiencies of non-templated addition.[33]

8.2.7 Overcoming Mis-Priming Events

Mis-priming events, resulting in the generation of non-specific amplification products, not only reduce the specificity of the PCR but also affect the sensitivity, repeatability and reproducibility of the reaction. Mis-priming events and the occurrence of other artefacts often increase as more primer sets are added to the reaction. Several methods for avoiding or overcoming such problems have been developed. Commonly, an increase in the annealing temperature is used to reduce artefacts but this must be balanced against the reduced yield of desired PCR products. This is a particularly fine balance if the optimal annealing temperatures of the constituent PCRs are not the same for all targets.

Hot-start PCR is a useful technique for reducing primer dimer and other non-specific primer interactions.[34] A number of methods have been developed for hot-start PCR (see Chapter 5). Touchdown PCR has also been successfully used to reduce the occurrence of non-specific products.[35] The technique involves setting the annealing temperature of the reaction artificially high for the initial cycles and then gradually reducing the set temperature over the following cycles to the empirically determined optimum. The initial high temperature may limit the generation of product but it ensures that the products generated are the desired ones. These products then act as templates in subsequent cycles. Reducing the annealing temperature improves the efficiency of the reaction and, given the increased concentration of the correct target, the occurrence of non-specific products is greatly reduced.

8.2.8 Specificity

The specificity of an mPCR may be improved by the following:

- Screening primers/probes to minimise possible homologies to other sequences and identification of possible secondary structures;

- Reducing primer and probe length by using base analogues to increase T_m;
- Use of different DNA polymerase enzymes or enzyme blends;
- Empirical optimisation of individual primer concentrations.

8.2.9 Untemplated Nucleotide Addition

A well-documented feature of *Taq* DNA polymerase is its tendency to add an additional untemplated nucleotide, usually an adenine, to the 3′ end of PCR product. This can result in a mixture of products for any one locus, some of which have an additional nucleotide and are therefore one base longer than products without the addition. This can cause difficulties in interpretation, especially in techniques such as the STR systems employed in forensic analysis, where single base-pair resolution of products is required.

In multiplex systems, which are necessarily carried out under conditions which are sub-optimal for some loci, but a working compromise for all, it is easier to optimise for the addition of the extra nucleotide, rather than its non-addition.

8.3 Detection Strategies

Commonly used detection strategies for mPCR are essentially the same as those for PCR, and the reader is referred to Chapter 5 for information.

8.4 Applications of mPCR

Multiplex PCR is an extremely flexible technique and has numerous applications, some of which are outlined in Table 8.2.

8.5 Advantages and Disadvantages of mPCR

Table 8.3 outlines some of the advantages and disadvantages of mPCR. Multiplex PCR is a valuable tool utilised in a wide range of applications from research and diagnostic PCR to forensic and environmental applications. Use of a multiplexed assay can save significant amounts of both time and resources for the analyst and the laboratory as a whole.

Here, the concepts of multiplex PCR have been introduced and some of the considerations that must be made when designing a successful mPCR system have been reviewed. In summary, two points should be stressed. First, the importance of good primer design. Whilst it may be tempting simply to take published primer sequences and combine them, such a policy is likely to give a sub-optimal and unstable multiplex system. Time spent to ensure that primer sets are compatible in terms of position and sequence is a sound investment.

Further to this is the importance of empirical optimisation. An optimised mPCR system cannot be achieved on paper. It will always be necessary to

Table 8.2 Applications of mPCR.

Application of mPCR	References
SNP detection: Using locus specific primers with distinct 5′-sequence tags (generic tags), genome-wide SNP detection has been made possible by combining highly multiplexed PCR reactions and a variety of detection systems	4,7,36–39
Pathogen identification	40,41
Gender screening	42,43
Forensic studies	4,44–47
Clinical/genetic research: A) Detection of chromosomal translocations in patients with acute myeloid or lymphoid leukaemia	A) 48
B) Screening for mutations in the CFTR gene potentially leading to cystic fibrosis, an example of multiplex allele-specific PCR used in genotyping	B) 49
C) Genetic disease diagnosis	C) 50–54
Sequencing	8,9,55,56

Table 8.3 Advantages and disadvantages of mPCR.

Advantages	Disadvantages
Saving time	Initial effort required in both theoretical and experimental design
Saving effort	The more primer sets added to the reaction the more difficult it becomes to ensure effective amplification and differentiation of all DNA fragments
Decreased reagent costs if using conventional primers	Increased reagent cost if using several primer sets each labelled with a fluorescent dye
Removal of tube-to-tube variation when using amplification controls	Detection equipment may be expensive, may require fluorescent detection systems or capillary electrophoresis to achieve sufficient size resolution
Can indicate template quality, when used in conjunction with target mimics[57]	Poor template quality may introduce bias

consider individual parameters and to optimise these by practical experimentation, within the framework of the developing system. However, it is also important to weigh the time spent on optimisation against the final demands on the system; a diagnostic service processing thousands of samples may justify greater resources given to optimisation than a small research project.

8.6 Introduction to WGA

In many areas of research such as forensic analysis, gene discovery and clinical research the paucity of genomic DNA available for analysis can be the major limiting factor determining the type and quantity of genetic testing that can be performed. One technology designed to overcome this problem is whole genome amplification (WGA), which was first developed in 1992.[58,59] The objective of this method is to amplify a limited DNA sample, in a sequence-independent manner, to generate a new sample that is indistinguishable from the original sample but with a higher DNA concentration.

This technique has been used to amplify DNA from a single cell[60] and from nanogram quantities of genomic DNA.[61] WGA methods have been utilised in many areas of research including analysis of human cancers,[62] pre-implantation genetic diagnosis[63] and prenatal diagnosis.[64,65] It is important that WGA methods produce an amplified un-biased representation of the entire genome, without the loss of any sequences that were within the original sample (allelic drop-out).

Various WGA techniques have been developed, which differ both in their protocols and in their replication accuracy. There are six common types of WGA methods:

- Degenerate oligonucleotide primed PCR (DOP-PCR);[59,61]
- Primer extension pre-amplification (PEP);[58]
- Improved primer extension pre-amplification (I-PEP);[66]
- Multiple displacement amplification (MDA);[67,68]
- Ligation mediated PCR (LMP);[69]
- T7-based linear amplification of DNA (TLAD).[70]

Each method will be discussed further in the sections below but for a more in-depth review the reader is referred to 'Whole Genome Amplification (Method Express)'.[71]

8.7 WGA Methodologies

8.7.1 Degenerate Oligonucleotide Primed PCR

Degenerate oligonucleotide primed PCR was first described in 1992 by Telenius et al.[59] DOP-PCR uses *Taq* polymerase and semi-degenerate oligonucleotide primers that bind at a low annealing temperature at approximately one million sites in the human genome. The first cycles using low annealing temperature are followed by a large number of PCR cycles with a higher annealing temperature, hence only the fragments that were tagged in the first step are amplified. DOP-PCR generates fragments that are on average 400–500 bp in length, with a maximum size of 3 kb, although a method able to produce fragments up to 10 kb has been developed.[72] The quality and amount of genomic DNA available for DOP-PCR can affect the efficiency and uniformity of amplification.

Kittler et al.[72] demonstrated that starting with less than 1 ng genomic DNA decreases the specificity of the amplification and increases the chance of allelic dropout; however, some groups have shown success with lower starting amounts of genomic DNA.[73] Given the possibility of loss of sequences, care should be taken when interpreting the results of DOP-PCR, particularly for clinical applications.

8.7.2 Primer Extension Pre-amplification

Primer extension pre-amplification was first described by Zhang et al.[58] This technique utilises *Taq* polymerase and a 15 bp random primer that initially anneals at low stringency (37 °C). The temperature is then gradually increased (by $0.1\,°C\,s^{-1}$) to 50 °C, followed by a final extension step at 55 °C for 4 minutes. Although the PEP protocol has been improved in different ways (improved-PEP, I-PEP),[66] it still results in incomplete genome coverage, failing to amplify certain sequences such as repeats, induces an amplification bias of the order of 10^3 to 10^6,[63,68,74,75] and has a limited efficiency on very small samples (such as single cells). Moreover, the use of *Taq* polymerase limits the maximal product length to about 3 kb. Much like DOP-PCR, PEP generates fragments on average of 400–500 bp. However, PEP has been demonstrated to be successful when starting from archival tissue that has either been formalin or ethanol fixed then paraffin embedded.[76–78]

8.7.3 Improved Primer Extension Pre-amplification

I-PEP is an improved PEP protocol that uses proof-reading enzymes and modified PCR conditions.[66,76] Compared to PEP, I-PEP has increased amplification efficiency, especially from low numbers of cells (1–5).

8.7.4 Multiple Displacement Amplification

MDA[79] is essentially an isothermal process, with a short initial denaturation step (94 °C, 2–3 min), a long amplification step (enzyme specific temperature, 6–18 h) and a short final enzyme inactivation step (65 °C, 15 min). The amplification process is based on the rolling circle amplification (RCA) type mechanism by which circular DNA molecules frequently replicate.[80] Due to the fact that random primers are used in MDA, sequence information on the target DNA is unnecessary. MDA uses either a combination of *Bst* polymerase (which amplifies at 50 °C) and T4 gene 32 protein, or *phi29* DNA polymerase (which amplifies at 30 °C) to amplify an entire genome. Products in excess of 10 kb can be amplified using this method.[68,81]

Template DNA is first denatured, the single-stranded DNA then acts as a template for random primer annealing and subsequent extension of the growing DNA chain by thousands of bases. The 5′ end of an extending strand is displaced by another upstream strand growing in the same direction

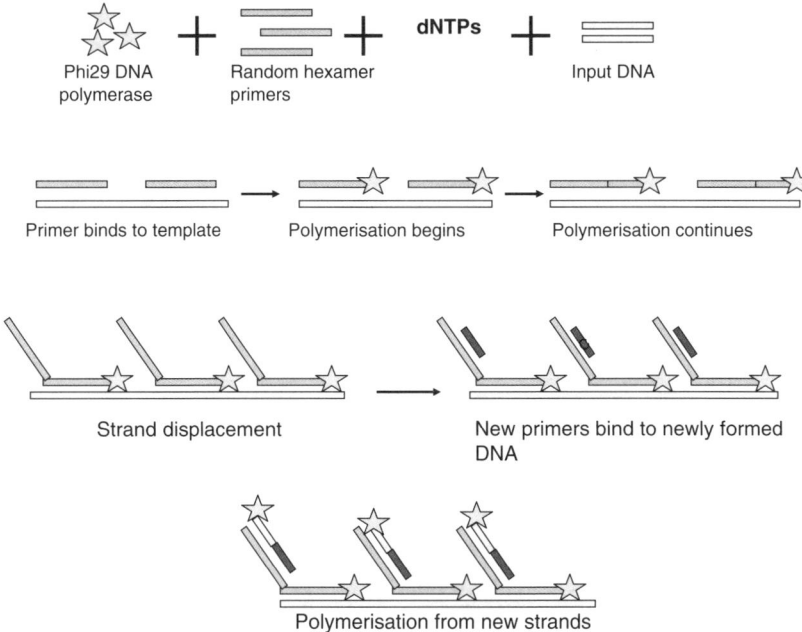

Figure 8.4 Schematic diagram of the amplification process for MDA. During MDA, random primers are annealed to the denatured DNA template and a DNA polymerase (*PHI29*) extends the growing DNA chain by thousands of bases. The 5′ end of an extending strand is displaced by another upstream strand growing in the same direction. This displaced DNA then acts as a template for new primers and hence a branched amplification mechanism is achieved.

(Figure 8.4). Due to the hyperbranching mechanism of MDA, microgram quantities of DNA can be generated in a few hours from as little as 5 ng of starting material.

8.7.5 Ligation-mediated PCR

Ligation-mediated PCR (LMP) was first described by Ludecke and coworkers[82] and was later adapted for the WGA of small quantities of genomic DNA[83] and for single cell comparative genomic hybridisation (SCOMP).[84] The LMP method uses endonuclease or chemical cleavage to fragment the genomic DNA sample; linkers are then ligated to the fragmented DNA, which serve as primers for amplification. The advantage of LMP is that it is a single-tube process, thus excessive sample handling and template loss is minimised (see Figure 8.5).

8.7.6 T7-based Linear Amplification of DNA

TLAD is a variant on the protocol originally designed by Phillips and Eberwine to amplify mRNA[70] that has been adapted for WGA.[85] The restriction enzyme

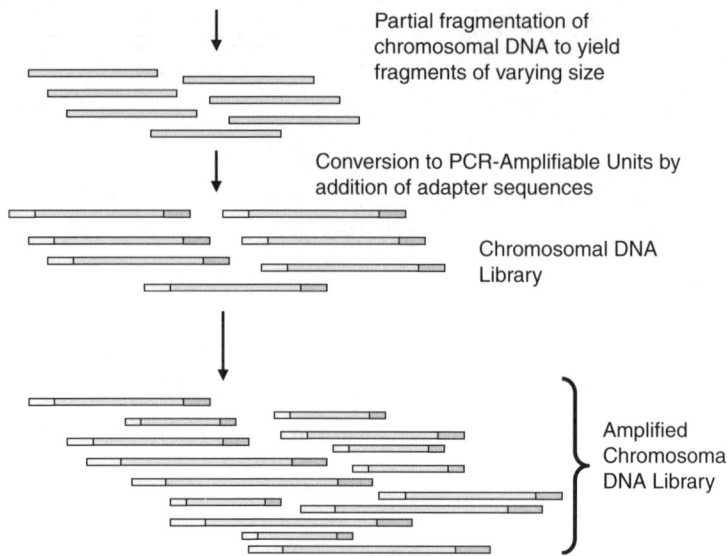

Figure 8.5 Schematic diagram of the amplification process for LMP. During LMP an endonuclease or chemical cleavage is used to first fragment the genomic DNA sample. Adapter sequences are then ligated to the DNA fragments, which serve as priming sites for PCR, resulting in thousands of copies of the original entire genomic DNA sample.

Alu I is used to digest the DNA, and terminal transferase is used to add a polyT tail on the 3' terminus. A primer containing a 5' T7 promoter and a 3' polyA tract is annealed to the DNA fragments. The second strand is synthesised using the Klenow fragment of DNA polymerase I, and the reaction products then serve as templates for the *in vitro* transcription reaction. The major advantage of TLAD is that it does not introduce sequence- and length-dependent biases. The major disadvantage is that the method is laborious.

8.8 WGA Applications and Characteristics

Some of the advantages, disadvantages and applications of the main WGA methods are outlined in Table 8.4.

Whole Genome Amplification is a powerful tool for increasing the amount of a limited DNA sample, facilitating subsequent analysis of the sample. Here the concepts and main approaches to WGA have been outlined, together with the benefits and drawbacks of the different methodologies available. Commercial kits are available for most WGA methods from a range of suppliers, which have the benefit of potentially reducing the time required in method optimisation and development. However, such kits are likely to have higher associated costs than in-house methods, which may preclude their use for some laboratories.

Table 8.4 Applications of WGA.

Method	Advantages	Disadvantages	Applications
DOP-PCR	Technically straightforward; Widely accepted; Commercial kits available	Amount of starting template can result in allelic drop out	SNP genotyping;[86,61,87] Array CGH;[88] Microsatellite genotyping[61]
PEP	Technically straightforward; Increased amplification efficiency compared to DOP-PCR; Commercial kits available	Incomplete genome coverage; Limited product length due to use of *Taq*	More suitable for archived tissues that have been paraffin embedded;[76,78] LOH testing in cancer diagnostics.[89]
I-PEP	Improved genome coverage compared to PEP; Utilises proof-reading enzyme resulting in increased sequencing accuracy; Commercial kits available	Lower yield compared to MDA	LOH testing;[90] Sequencing;[66] Microsatellite analysis[91]
MDA	Improved genome coverage compared to DOP-PCR and PEP; Results in microgram quantities of genomic DNA; Commercial kits available	Long reaction time; Less specific compared to I-PEP	SNP analysis;[87,92] qPCR;[93] Array CGH[68]
LMP	Use of single primer for all fragments; Single-tube process; Commercial kits available	Technically difficult	Microsatellite genotyping;[94] Sequence analysis; Array CGH
TLAD	Results in microgram quantities of genomic DNA; Less amplification bias compared to PCR-based methods	Multiple steps involved, requiring sample clean-up at each stage; Time consuming; No commercial kit available	Biomarker screening[95] cDNA gene expression analysis[96]

The results of analyses performed on amplified material should be carefully scrutinised to ensure that amplification bias or allelic drop-out have not influenced the results obtained. A recent assessment of several WGA approaches using whole genome sequencing to determine the integrity and uniformity of amplification of two bacterial genomes demonstrated that all the methods tested introduced some bias.[97] In addition, when choosing a WGA method consideration should be given to the amount and quality of starting material available and how it has been processed, as this can affect the genome coverage of the chosen method. The yield and product length of the WGA method should also be chosen to match the intended application.

Acknowledgements

The authors would like to acknowledge Jim Thomson and Jo Short for the production of material that has been used in writing this chapter and David French for helpful discussions about multiplex PCR.

References

1. J. S. Chamberlain, R. A. Gibbs, J. E. Ranier, P. N. Nguyen and C. T. Caskey, *Nucleic Acids Res.*, 1988, **16**, 11141.
2. J. S. Chamberlain and J. R. Chamberlain, in *PCR Polymerase Chain Reaction*, ed. K. B. Mullis, F. Ferre and R. A. Gibbs, Birkhäuser, Boston, 1994, p. 38.
3. B. M. Pearson and R. A. McKee, *Int. J. Food Microbiol.*, 1992, **16**, 63.
4. S. Inagaki, Y. Yamamoto, Y. Doi, T. Takata, T. Ishikawa, K. Imabayashi, K. Yoshitome, S. Miyaishi and H. Ishizu, *Forensic Sci. Int.*, 2004, **144**, 45.
5. L. Li, C. T. Li, R. Y. Li, Y. Liu, Y. Lin, T. Z. Que, M. Q. Sun and Y. Li, *Forensic Sci. Int.*, 2006, **162**, 74.
6. J. Xu, H. Miao, H. Wu, W. Huang, R. Tang, M. Qiu, J. Wen, S. Zhu and Y. Li, *Biosens. Bioelectron.*, 2006, **22**, 71.
7. J. Rachlin, C. Ding, C. Cantor and S. Kasif, *BMC Genomics*, 2005, **6**, 102.
8. G. M. van der Vliet, M. Y. de Wit and P. R. Klatser, *Mol. Cell Probes*, 1993, **7**, 61.
9. H. Cuppens, I. Buyse, M. Baens, P. Marynen and J. J. Cassiman, *Mol. Cell Probes*, 1992, **6**, 33.
10. D. G. Wang, J. B. Fan, C. J. Siao, A. Berno, P. Young, R. Sapolsky, G. Ghandour, N. Perkins, E. Winchester, J. Spencer, L. Kruglyak, L. Stein, L. Hsie, T. Topaloglou, E. Hubbell, E. Robinson, M. Mittmann, M. S. Morris, N. Shen, D. Kilburn, J. Rioux, C. Nusbaum, S. Rozen, T. J. Hudson, R. Lipshutz, M. Chee and E. S. Lander, *Science*, 1998, **280**, 1077.
11. J. G. Hacia, B. Sun, N. Hunt, K. Edgemon, D. Mosbrook, C. Robbins, S. P. Fodor, D. A. Tagle and F. S. Collins, *Genome Res.*, 1998, **8**, 1245.
12. O. Henegariu, N. A. Heerema, S. R. Dlouhy, G. H. Vance and P. H. Vogt, *Biotechniques*, 1997, **23**, 504.

13. http://info.med.yale.edu/genetics/ward/tavi/Guide.html (accessed April 2007).
14. http://info.med.yale.edu/genetics/ward/tavi/Trblesht.html (accessed April 2007).
15. http://ngrl.man.ac.uk/SNPCheck/dbSNP: (accessed April 2007).
16. http://www.ncbi.nlm.nih.gov/projects/SNP/Ensembl: (accessed April 2007).
17. http://www.ensembl.org/index.html (accessed April 2007).
18. http://www.dnasoftware.com/?gclid=CJmwvfDC5YoCFSYSQgodzAL73Q. (accessed March 2007).
19. https://products.appliedbiosystems.com. (accessed April 2007).
20. http://www.bioinfo.rpi.edu/applications/mfold/dna/form1.cgi (accessed April 2007).
21. Y. Lebedev, N. Akopyants, T. Azhikina, Y. Shevchenko, V. Potapov, D. Stecenko, D. Berg and E. Sverdlov, *Genet. Anal.*, 1996, **13**, 15.
22. H. K. Nguyen, E. Bonfils, P. Auffray, P. Costaglioli, P. Schmitt, U. Asseline, M. Durand, J. C. Maurizot, D. Dupret and N. T. Thuong, *Nucleic Acids Res.*, 1998, **26**, 4249.
23. http://www.glenres.com (accessed April 2007).
24. D. Loakes and D. M. Brown, *Nucleic Acids Res.*, 1994, **22**, 4039.
25. E. Ohtsuka, S. Matsuki, M. Ikehara, Y. Takahashi and K. Matsubara, *J. Biol. Chem.*, 1985, **260**, 2605.
26. A. Van Aerschot, J. Rozenski, D. Loakes, N. Pillet, G. Schepers and P. Herdewijn, *Nucleic Acids Res.*, 1995, **23**, 4363.
27. K. Y. Lin and J. Matteucci, *J. Am. Chem. Soc.*, 1998, **120**, 8531.
28. Z. Dogan, R. Paulini, J. A. Rojas Stutz, S. Narayanan and C. Richert, *J. Am. Chem. Soc.*, 2004, **126**, 4762.
29. S. Narayanan, J. Gall and C. Richert, *Nucleic Acids Res.*, 2004, **32**, 2901.
30. S. K. Singh, P. Nielsen, A. A. Koshkin and J. Wengel, *Chem. Commun. (Cambridge, UK)*, 1988, **4**, 455.
31. http://www.eurogentec.com/module/FileLib/oligo_16.pdf (accessed April 2007).
32. T. Takiya, Y. Seto, H. Yasuda, T. Suzuki and K. Kawai, *Nucleic Acids Symp. Ser. (Oxford, UK)*, 2004, **131**.
33. G. Hu, *DNA Cell Biol.*, 1993, **12**, 763.
34. R. T. D'Aquila, L. J. Bechtel, J. A. Videler, J. J. Eron, P. Gorczyca and J. C. Kaplan, *Nucleic Acids Res.*, 1991, **19**, 3749.
35. C. Mellersh and J. Sampson, *Biotechniques*, 1993, **15**, 582.
36. J. N. Hirschhorn, P. Sklar, K. Lindblad-Toh, Y. M. Lim, M. Ruiz-Gutierrez, S. Bolk, B. Langhorst, S. Schaffner and E. Winchester, *Proc. Natl. Acad. Sci. U.S.A.*, 2000, **97**, 12164.
37. J. J. Sanchez, C. Phillips, C. Borsting, K. Balogh, M. Bogus, M. Fondevila, C. D. Harrison, E. Musgrave-Brown, A. Salas, D. Syndercombe-Court, P. M. Schneider, A. Carracedo and N. Morling, *Electrophoresis*, 2006, **27**, 1713.
38. P. A. Orlandi, L. Carter, A. M. Brinker, A. J. Da Silva, D. M. Chu, K. A. Lampel and S. R. Monday, *Appl. Environ. Microbiol.*, 2003, **69**, 4806.
39. M. M. Shi, *Clin. Chem. (Washington, DC)*, 2001, **47**, 164.

40. E. Sanjuan and C. Amaro, *Appl. Environ. Microbiol.*, 2007, **73**, 2029.
41. D. S. Billal, M. Hotomi, M. Suzumoto, K. Yamauchi, I. Kobayashi, K. Fujihara and N. Yamanaka, *Int. J. Pediatr. Otorhinolaryngol.*, 2007, **71**, 269.
42. E. Jauniaux, V. Cirigliano and M. Adinolfi, *Reprod. Biomed. Online*, 2003, **6**, 494.
43. G. Levinson, R. A. Fields, G. L. Harton, F. T. Palmer, A. Maddalena, E. F. Fugger and J. D. Schulman, *Hum. Reprod.*, 1992, **7**, 1304.
44. T. M. Clayton, J. P. Whitaker, R. Sparkes and P. Gill, *Forensic Sci. Int.*, 1998, **91**, 55.
45. H. Asamura, S. Fujimori, M. Ota and H. Fukushima, *Forensic Sci. Int.*, 2007, **173**, 7.
46. C. P. Kimpton, N. J. Oldroyd, S. K. Watson, R. R. Frazier, P. E. Johnson, E. S. Millican, A. Urquhart, B. L. Sparkes and P. Gill, *Electrophoresis*, 1996, **17**, 1283.
47. K. L. Swango, W. R. Hudlow, M. D. Timken and M. R. Buoncristiani, *Forensic Sci. Int.*, 2007, **170**, 33.
48. N. Pallisgaard, P. Hokland, D. C. Riishoj, B. Pedersen and P. Jorgensen, *Blood*, 1998, **92**, 574.
49. http://www.elucigene.co.uk/(accessed March 2007).
50. T. A. Cebula, W. L. Payne and P. Feng, *J. Clin. Microbiol.*, 1995, **33**, 248.
51. R. D. Klein, E. C. Thorland, P. R. Gonzales, P. A. Beck, D. J. Dykas, J. M. McGrath and A. E. Bale, *Clin. Chem. (Washington, DC)*, 2006, **52**, 1864.
52. P. Renwick and C. M. Ogilvie, *Expert Rev. Mol. Diagn.*, 2007, **7**, 33.
53. P. C. Patsalis, L. Kousoulidou, K. Mannik, C. Sismani, O. Zilina, S. Parkel, H. Puusepp, N. Tonisson, P. Palta, M. Remm and A. Kurg, *Eur. J. Hum. Genet.*, 2007, **15**, 162.
54. Q. R. Chen, G. Vansant, K. Oades, M. Pickering, J. S. Wei, Y. K. Song, J. Monforte and J. Khan, *J. Mol. Diagn.*, 2007, **9**, 80.
55. M. Spentchian, I. Brun-Heath, A. Taillandier, D. Fauvert, J. L. Serre, B. Simon-Bouy, F. Carvalho, I. Grochova, S. G. Mehta, G. Muller, S. L. Oberstein, G. Ogur, S. Sharif and E. Mornet, *Genet. Test.*, 2006, **10**, 252.
56. Z. Sun, L. Zhou, H. Zeng, Z. Chen and H. Zhu, *Genomics*, 2007, **89**, 151.
57. G. Gilliland, S. Perrin and H. Bunn, in *PCR Protocols: a Guide to Methods and Applications*, ed. M. A. Innis, D. H. Gelfand, J. J. Sninsky and T. J. White, Academic Press, San Diego, CA, 1990, p. 69.
58. L. Zhang, X. Cui, K. Schmitt, R. Hubert, W. Navidi and N. L. Arnheim, *Proc. Natl. Acad. Sci. U.S.A.*, 1992, **89**, 5847.
59. H. Telenius, N. P. Carter, C. E. Bebb, M. Nordenskjold, B. A. Ponder and A. H. Tunnacliffe, *Genomics*, 1992, **13**, 718.
60. M. C. Snabes, S. S. Chong, S. B. Subramanian, K. Kristjansson, D. DiSepio and M. R. Hughes, *Proc. Natl. Acad. Sci. U.S.A.*, 1994, **91**, 6181.
61. V. G. Cheung and S. F. Nelson, *Proc. Natl. Acad. Sci. U.S.A.*, 1996, **93**, 14676.

62. A. Ao, D. Wells, A. H. Handyside, R. M. Winston and J. D. Delhanty, *J. Assist. Reprod. Genet.*, 1998, **15**, 140.
63. K. Sermon, W. Lissens, H. Joris, A. Van Steirteghem and I. Liebaers, *Mol. Hum. Reprod.*, 1996, **2**, 209.
64. A. Sekizawa, T. Kimura, M. Sasaki, S. Nakamura, R. Kobayashi and T. Sato, *Neurology*, 1996, **46**, 1350.
65. A. Sekizawa, A. Watanabe, T. Kimura, H. Saito, T. Yanaihara and T. Sato, *Obstet. Gynecol.*, 1996, **87**, 501.
66. W. Dietmaier, A. Hartmann, S. Wallinger, E. Heinmoller, T. Kerner, E. Endl, K. W. Jauch, F. Hofstadter and J. Ruschoff, *Am. J. Pathol.*, 1999, **154**, 83.
67. F. B. Dean, J. R. Nelson, T. L. Giesler and R. S. Lasken, *Genome Res.*, 2001, **11**, 1095.
68. F. B. Dean, S. Hosono, L. Fang, X. Wu, A. F. Faruqi, P. Bray-Ward, Z. Sun, Q. Zong, Y. Du, J. Du, M. Driscoll, W. Song, S. F. Kingsmore, M. Egholm and R. S. Lasken, *Proc. Natl. Acad. Sci. U.S.A.*, 2002, **99**, 5261.
69. R. Lucito, M. Nakimura, J. A. West, Y. Han, K. Chin, K. Jensen, R. McCombie, J. W. Gray and M. Wigler, *Proc. Natl. Acad. Sci. U.S.A.*, 1998, **95**, 4487.
70. J. Phillips and J. H. Eberwine, *Methods (Oxford, UK)*, 1996, **10**, 283.
71. S. Hughes and R. S. Lasken, *Whole Genome Amplification*, Scion Publishing Ltd, Oxford, UK, 2005. ISBN 1-904842-07-0.
72. R. Kittler, M. Stoneking and M. Kayser, *Anal. Biochem.*, 2002, **300**, 237.
73. Q. Huang, S. P. Schantz, P. H. Rao, J. Mo, S. A. McCormick and R. S. Chaganti, *Gene. Chromosome Canc.*, 2000, **28**, 395.
74. K. Xu, Y. Tang, J. A. Grifo, Z. Rosenwaks and J. Cohen, *Hum. Reprod.*, 1993, **8**, 2206.
75. Z. Jiao, C. Zhou, J. Li, Y. Shu, X. Liang, M. Zhang and G. Zhuang, *Prenat. Diagn.*, 2003, **23**, 646.
76. E. Heinmoller, Q. Liu, Y. Sun, G. Schlake, K. A. Hill, L. M. Weiss and S. S. Sommer, *Lab. Invest.*, 2002, **82**, 443.
77. E. Heinmoller, A. Bockholt, M. Werther, M. Ziemer, A. Muller, B. M. Ghadimi and J. Ruschoff, *Pathol. Res. Pract.*, 2003, **199**, 363.
78. F. Bataille, P. Rummele, W. Dietmaier, D. Gaag, F. Klebl, A. Reichle, P. Wild, F. Hofstadter and A. Hartmann, *Mol. Pathol.*, 2003, **56**, 286.
79. P. M. Lizardi, 2000, US Patent 6,123,120.
80. A. Kornberg and T. A. Baker, *Deoxyribonucleic Acid Replication*, W. H. Freeman and Company, San Francisco, 1992.
81. J. M. Lage, J. H. Leamon, T. Pejovic, S. Hamann, M. Lacey, D. Dillon, R. Segraves, B. Vossbrinck, A. Gonzalez, D. Pinkel, D. G. Albertson, J. Costa and P. M. Lizardi, *Genome Res.*, 2003, **13**, 294.
82. H. J. Ludecke, G. Senger, U. Claussen and B. Horsthemke, *Nature*, 1989, **338**, 348.
83. C. Tanabe, K. Aoyagi, T. Sakiyama, T. Kohno, N. Yanagitani, S. Akimoto, M. Sakamoto, H. Sakamoto, J. Yokota, M. Ohki, M. Terada, T. Yoshida and H. Sasaki, *Gene. Chromosome Canc.*, 2003, **38**, 168.

84. C. A. Klein, O. Schmidt-Kittler, J. A. Schardt, K. Pantel, M. R. Speicher and G. Riethmuller, *Proc. Natl. Acad. Sci. U.S.A.*, 1999, **96**, 4494.
85. C. L. Liu, S. L. Schreiber and B. E. Bernstein, *BMC Genomics*, 2003, **4**, 19.
86. S. F. Grant, S. Steinlicht, U. Nentwich, R. Kern, B. Burwinkel and R. Tolle, *Nucleic Acids Res.*, 2002, **30**, e125.
87. D. L. Barker, M. S. Hansen, A. F. Faruqi, D. Giannola, O. R. Irsula, R. S. Lasken, M. Latterich, V. Makarov, A. Oliphant, J. H. Pinter, R. Shen, I. Sleptsova, W. Ziehler and E. Lai, *Genome Res.*, 2004, **14**, 901.
88. Y. J. Chung, J. Jonkers, H. Kitson, H. Fiegler, S. Humphray, C. Scott, S. Hunt, Y. Yu, I. Nishijima, A. Velds, H. Holstege, N. Carter and A. Bradley, *Genome Res.*, 2004, **14**, 188.
89. D. J. Simpson, E. J. Bicknell, H. N. Buch, S. J. Cutty, R. N. Clayton and W. E. Farrell, *Gene. Chromosome Canc.*, 2003, **37**, 225.
90. J. M. van Oers, C. Adam, S. Denzinger, R. Stoehr, S. Bertz, D. Zaak, C. Stief, F. Hofstaedter, E. C. Zwarthoff, T. H. van der Kwast, R. Knuechel and A. Hartmann, *Int. J. Cancer*, 2006, **119**, 1212.
91. P. J. Wild, R. Stoehr, R. Knuechel, A. Hartmann and W. Dietmaier, *Methods Mol. Biol.*, 2005, **293**, 93.
92. K. K. Wong, Y. T. Tsang, J. Shen, R. S. Cheng, Y. M. Chang, T. K. Man and C. C. Lau, *Nucleic Acids Res.*, 2004, **32**, e69.
93. S. Hosono, A. F. Faruqi, F. B. Dean, Y. Du, Z. Sun, X. Wu, J. Du, S. F. Kingsmore, M. Egholm and R. S. Lasken, *Genome Res.*, 2003, **13**, 954.
94. J. Li, L. Harris, H. Mamon, M. H. Kulke, W. H. Liu, P. Zhu and G. Makrigiorgos, *J. Mol. Diagn.*, 2006, **8**, 22.
95. B. K. Petroff, T. A. Phillips, B. F. Kimler and C. J. Fabian, *Anal. Quant. Cytol. Histol.*, 2006, **28**, 297.
96. N. Mutsuga, T. Shahar, J. G. Verbalis, C. C. Xiang, M. J. Brownstein and H. Gainer, *Endocrinology*, 2005, **146**, 1254.
97. R. Pinard, A. de Winter, G. J. Sarkis, M. B. Gerstein, K. R. Tartaro, R. N. Plant, M. Egholm, J. M. Rothberg and J. H. Leamon, *BMC Genomics*, 2006, **7**, 216.

CHAPTER 9
Procedures for Quality Control of RNA Samples for Use in Quantitative Reverse Transcription PCR

TANIA NOLAN[1,2] AND STEPHEN BUSTIN[2,3]

[1] Sigma Aldrich, Cambridge UK; [2] Eureka Biotechnology, Cambridge UK; [3] Institute of Cell and Molecular Science, Barts and the London Queen Mary's School of Medicine and Dentistry, London, UK

9.1 Introduction

The quality of any scientific data is directly proportional to that of the original starting samples, or simply 'garbage in, garbage out'. In most circumstances it is logical to work with the highest quality material possible. However, for some experiments the highest quality possible is still a serious compromise from perfection. The degree to which the standard of input material influences final quantitative reverse transcription polymerase chain reaction (qRT-PCR) data and, potentially, the resulting scientific conclusion, is outlined in this chapter.

9.2 RNA Extraction Approaches

In order to have the best possible chance of extracting high-quality RNA, tissue and cell samples should be extracted from the source and RNAse activity prevented as quickly as possible.

9.2.1 Freezing

Solid tissue biopsies need to be stabilised immediately at source with subsequent RNA extraction procedures carried out when required. This can be achieved by snap-freezing in liquid nitrogen. For these samples, labelling and cataloguing must be rigorous to ensure rapid and accurate retrieval and that tube markings are not removed during freezing. Automated storage and retrieval systems have revolutionised the whole process of sample tracking.

9.2.2 Sulfate

An alternative to freezing is to immerse tissues into aqueous sulfate salt solutions (such as ammonium sulfate) at controlled pH and ambient temperature. This treatment results in precipitation of RNases and other solubilised proteins and protects tissue RNA. Tissue samples should be prepared as slices less than 0.5 cm, preferably 2 mm.[1] A larger relative surface area facilitates diffusion of the solution into the tissue. Treated tissues can be stored at –60 °C prior to processing using standard RNA preparation techniques.[2] This technique forms the basis of the commercially available RNA*later*® solution (Ambion Inc., Applied Biosystems, USA). Small organs such as rat livers or kidneys can be immersed whole in solution, small sections of tissue less than 0.5 cm thick should be stored in 5 volumes RNA*later*®. Cell culture pellets can be re-suspended in a minimal volume of phosphate buffered saline (PBS) and then 5–10 volumes RNA*later*® added. Samples can be stored in RNA*later*® at ambient temperature for up to 1 week, or long term at –20 °C. RNA*later*-ICE® (Ambion Inc., Applied Biosystems, USA) has been developed to aid tissue processing of previously frozen material. These samples are then processed using conventional column or phenol based systems such as TRI®Reagent (Sigma Aldrich, USA).

9.2.3 Guanidinium Isothiocyanate

It is preferable to harvest adherent cultured cells directly in lysis buffer containing guanidinium isothiocyanate. This process enables maintenance of representative cellular messages because it ensures rapid inactivation of RNases that are released during trypsin treatment and can subsequently initiate mRNA degradation. Similarly, small tissue sections can be homogenised directly in guanidinium isothiocyanate lysis buffer.

9.2.4 Phenol

Alternatively cells or tissue can be disrupted in TRI®Reagent. These homogenates can also be stored at –80 °C until RNA purification is required. RNA extracted using these phenol-based protocols results in a high yield of nucleic acid, but care must be exercised to ensure high levels of purity.

9.2.5 Additional Purification

In some cases it is appropriate to perform a subsequent column purification step and DNase I digestion to ensure removal of protein and genomic DNA (gDNA) contamination. Column-based purification procedures in kit format usually produce pure RNA samples. In most cases a gDNA removal procedure is incorporated into the protocol. Performing this reaction *via* a column ensures that residual gDNA or any DNase I activity does not remain in the sample.

9.2.6 Extraction from Archival Tissue Samples

Archived formalin fixed, paraffin embedded tissue (FFPE) samples have been explored as a rich source of RNA from samples with complete histological profiles. These samples offer the potential to investigate a number of disorders because they are usually accompanied by detailed medical histories and clinical outcomes.[3] RNA from fixed tissues is usually more difficult to extract due to cross-linking to proteins[4] and the fixation and storage process often results in RNA degradation.[5] The fixing process also results in mono-methylol modifications on all bases, which results in inhibition of subsequent reverse transcription reactions.

It is for these reasons that RNA extracted from formalin fixed material is invariably low quality and can produce results that deviate from those derived from fresh tissues. In an investigation into the potential influence of sample processing and storage, tissue sections were divided and sections either frozen or formalin fixed. RNA was extracted from each section and the copy number of specific mRNA targets was determined using gene-specific reverse transcription and reference to a calibration curve constructed from an artificial oligonucleotide. The hypothesis was that the proportion of transcripts detected in fresh tissue relative to formalin fixed tissue would remain constant if the fixation procedure affected all tissues equally, indicating that fixed material could be used as a reliable source of RNA for qRT-PCR determination of gene quantification. The initial observation was that there was an increase in the variability of quantities detected in the replicate tissues sections after formalin fixation when compared to the reproducibility in quantities measured in samples after freezing. There was a six-fold relative difference in the quantity of glyceraldehyde-3-phosphate dehydrogenase (GAPDH) extracted from frozen and fixed tissues. However, this difference was not consistent for other transcript quantities; there was a five-fold difference in the quantity of vitamin D receptor (VDR) between the two tissue treatments, a ten-fold difference in insulin-like growth factor I receptor (IGF-IR) and yet only a two-fold difference in 24-O Hydroxylase (24-OHase). The variability in the observed differences demonstrates that sample freezing and formalin fixation result in inconsistencies in transcript quantification, potentially resulting in different biological conclusions[6] (Figure 9.1).

In an extensive study of the factors influencing qRT-PCR of extracts from FFPE material, Godfrey *et al.*[7] demonstrated that the highest quality RNA was produced after two sequential TRIzoL® (Invitrogen, UK) extractions, targets were more efficiently detected with amplicons < 130 bp and careful optimisation of RT conditions were required. Despite optimisation improving the data from fixed tissue the authors note that detection of targets is less efficient from

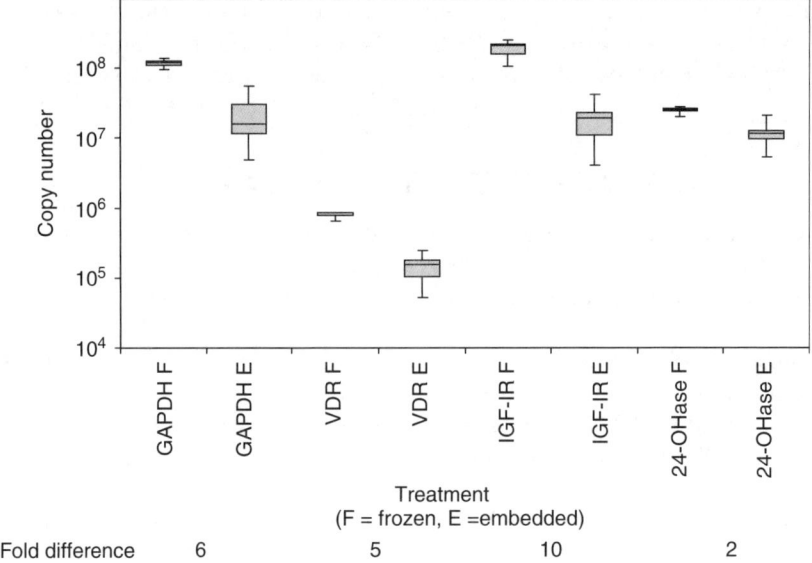

Figure 9.1 Graph showing the amount of RNA quantified from frozen (F) formalin fixed embedded (E) tissue samples. The fold difference in RNA detected for four transcripts is shown; glyceraldehyde-3-phosphate dehydrogenase (GAPDH), vitamin D receptor (VDR), insulin-like growth factor I receptor (IGF-IR), 24-O Hydroxylase (24-OHase).

fixed material and that the effect on different mRNA species, and even different fragments of the same mRNA, is variable[8] (also shown in Figure 9.1). Interestingly these authors report that pre-fixation time had the least effect on mRNA quantification but Macabeo-Ong et al.[9] report that prolonged formalin fixation had a detrimental effect on qRT-PCR. A further technical problem is highlighted by Williams et al.,[10] who demonstrated that in RNA extracted from FFPE tissue, as many as 1:500 bases are mutated. These base changes are either C to T or G to A transitions. These data indicate that quantification of mRNA from formalin fixed tissue must be carried out with great care and with the knowledge that relative transcript quantities may not be accurate.

A relatively new detection approach, the QuantiGene® branched DNA detection method (Panomics Inc., USA), may be more appropriate than qRT-PCR for the detection of damaged and chemically modified material.[11,12]

9.3 RNA Quality

Since tissue storage and treatment and RNA extraction procedures are so variable it is imperative that a reliable protocol for analysis of sample quality and quantity is defined. A full description of an RNA sample requires a statement regarding quality and a measure of quantity. RNA quality is a factor of both the purity of the sample and the degradation status of the RNA molecules. When the sample is to be used for measurements of transcript quantity one relevant measurement is a determination of whether the mRNA molecules are degraded.

Traditionally, analysis of RNA quality was by gel electrophoresis and analysis of the ratio of the quantities of the ribosomal RNA molecules. Using the ratio of the ribosomal fragments is unreliable because it relies on transcript independent molecules to infer the mRNA status.

9.3.1 RNA Integrity Number

In a recent report, Schroeder et al.[13] suggested that it is possible to calculate a more objective measure of RNA quality by measuring characteristics of the electropherogram generated by the Agilent 2100 Bioanalyzer, including the fraction of the area in the region of 18S and 28S rRNA, the height of the 28S peak, the presence or absence of RNA degradation products, the fast area ratio and marker height. These features were used to calculate and assign an RNA Integrity Number (RIN) to each RNA sample. RIN values range from 1 for completely degraded samples to a value of 10 for completely intact RNA. However, in an elegant study to investigate the influence of RNA integrity on qRT-PCR assay performance, Fleige and Pfaffl[14] reach a different conclusion. The authors extracted RNA samples from numerous bovine tissue types, subjected them to controlled degradation and analysed them using the Agilent 2100 Bioanalyzer. The samples had RINs between 10, which were apparently intact, to 4 with almost no evidence of rRNA bands. The quantity of individual transcripts in each of these samples was then determined using qRT-PCR assays. In some tissues the quantity of the measured transcripts was independent of RIN whereas in others there was a linear relationship and in still others a threshold response. Critically, the relationship between transcript quantity and RIN was different for different tissues and different transcripts and there was not a predictable relationship between these factors. The authors conclude that moderately degraded RNA samples can be reliably analysed and quantified using short amplicons (<250 bp) and expression is normalised against an internal reference and recommend that a RIN of at least 8 is required to assume that RNA is high quality. Similarly, in an evaluation of the stability of reference gene transcripts in experimental samples Pérez-Novo et al.[15] conclude that 'it is inappropriate to compare intact and degraded samples'. The discrepancy between the initial report describing the RIN algorithm and these evaluations of the correlation of RIN to transcript quantification[14,15] could be due to the relatively poor correlation coefficient (0.52) between RIN and expression values of the reference genes reported by the authors advocating the use of RINs.[13]

In the absence of an alternative reliable measure of mRNA integrity, the use of a 3′:5′ assay using GAPDH as the target sequence has been proposed.[16]

The data obtained are independent of ribosomal RNA integrity, provide a reasonable measure of the degradation of the transcripts of interest and are modelled on the standard approach adopted by microarray users and long-accepted conventional techniques applied to end-point PCR assays.[17] The 3′:5′ assay measures the integrity of the ubiquitously expressed mRNA specified by the GAPDH gene, which in this example is taken as representative of the integrity of all mRNAs in an RNA sample. However, since different mRNAs degrade at different rates, this may not always be the case and it may be necessary to design similar assays for specific targets. The RT reaction of the GAPDH mRNA is primed using oligo-dT, and a separate multiplex PCR assay is used to quantify the levels of three target amplicons. These are spatially separated with one towards the 5′ end, the second towards the centre and the third towards the 3′ end of the mRNA sequence. The ratio of amplicons reflects the relative success of the oligo-dT primed RT to proceed along the entire length of the transcript. This is prematurely terminated when mRNA is degraded. Consequently, a 3′:5′ ratio of around 1 indicates high integrity, whereas anything greater than 5 suggests degradation. The assay is designed as a triplex assay using TaqMan™ chemistry such that each amplicon is detected by a target-specific, differentially labelled probe. An example of the use of the 3′:5′ assay to evaluate RNA samples is shown in Figure 9.2.

The 3′:5′ assay is particularly applicable for analysis of precious samples when little RNA is available. An example of analysing RNA extracted from FFPE tissue is shown in Figure 9.2C. There are at least 4 Cts difference between the detection of each of the GAPDH assays indicating that this RNA is seriously degraded.

9.3.2 Spectrophotometric Measurement

A_{260}/A_{280} measurements are often made in an attempt to assess the quality of nucleic acid samples. These measurements are based upon the ratio between the absorbance of nucleic acid at A_{260} and protein and indicate absorbance of protein and phenol at A_{280}. A ratio below 1.8 generally indicates the presence of substances absorbing at A_{280} and usually the sample is considered to contain contamination. It is clear that this is not a reliable measure of sample quality since it is limited in the range of substances detected and does not reveal degradation state.

9.3.3 Presence of Inhibitors

Inhibitory components frequently found in biological samples can result in a significant reduction in the sensitivity and kinetics of qPCR.[18–23] The inhibiting agents may be reagents used during nucleic acid extraction or co-purified components from the biological sample, for example bile salts, urea, heme, heparin or IgG. The potential inaccuracies occur when an external calibration curve is used to calculate the number of transcripts in test samples. Invariably the material used to produce the calibration curve is biologically distinct from the test material, which is more likely to contain inhibitors. This leads to an underestimation of the mRNA levels in the test samples.[24] As discussed previously, the increasing interest in extracting nucleic acids from FFPE archival material will undoubtedly lead to an exacerbation of this problem.

The most common procedure used to account for any differences in PCR efficiencies between samples is to amplify a reference gene in parallel to the reporter gene and relate their quantification. However, this approach assumes that the two assays are inhibited to the same degree. In an attempt to demonstrate the effect of a contaminating agent in an RNA sample, EDTA was added to purified RNA samples to a final concentration of 125 mM. This sample was also included in the GAPDH 3′:5′ assay analysis described previously as an assay for detection of degraded RNA (Figure 9.3). A clear shift to higher Ct was observed for both the 3′ and 5′ GAPDH assays (relative to the

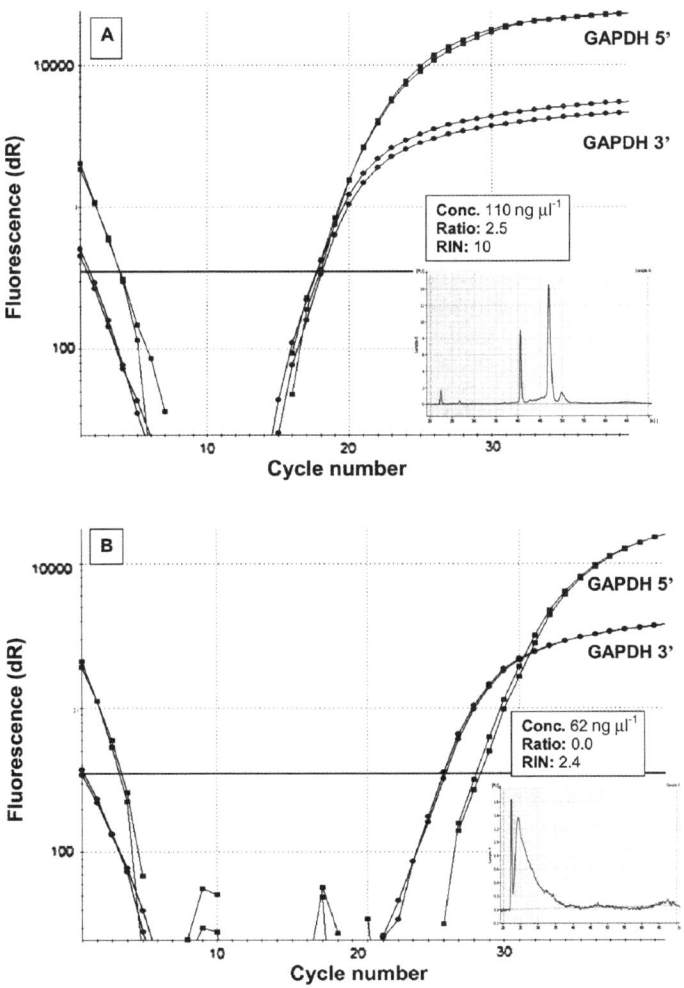

Figure 9.2 *(Continued)*

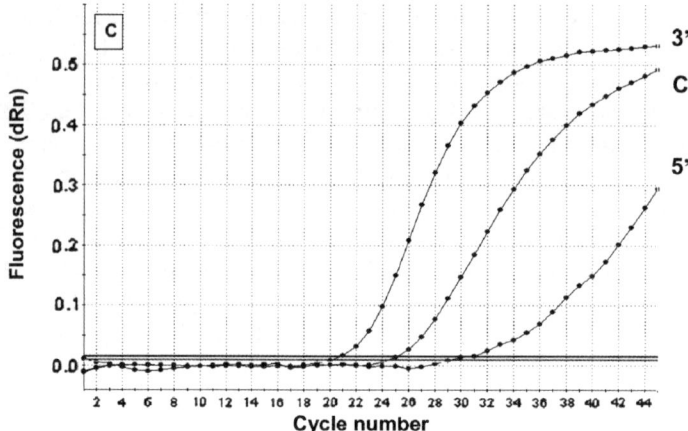

Figure 9.2 GAPDH mRNA quantified from oligo dT primed cDNA using three individual qPCR assays targeting 5′, centre and 3′ regions. (A) This sample has an Agilent 2100 Bioanalyzer RIN of 10 and equal concentrations of 5′ and 3′ assay target sequences. (B) The second sample appears to be seriously degraded, RIN 2.4, and has an apparently lower concentration. The 3′ GAPDH assay detects a higher concentration of target with Ct 24 than the 5′ with Ct 27 confirming degradation of this sample. Since the second sample was produced from the first it is important to note that the shift in Ct from 18 to 24 for the 3′ assay is indicative of the degree of degradation. (C) RNA extracted from FFPE tissue showing differences in the quantities of 5′, centre and 3′ sequences indicating that this RNA is seriously degraded.

pure RNA sample, Figure 9.2A). In the presence of the inhibitor both reactions are inhibited but the effect on the 5′ reaction is more pronounced than on the 3′. Since it can be assumed that reverse transcription of the 3′ site must precede that of the 5′ site, the higher yield of the 5′ target is due to greater sensitivity of the 3′ assay to the effect of EDTA inhibition. This is a single example that clearly indicates that qRT-PCR assays may be differentially affected by inhibitors. This demonstrates clearly that it is inappropriate to assume that the effect of inhibition is equal for all qPCR assays. Therefore, the presence of inhibitors cannot be cancelled by reference to a second target amplification or normalisation to a reference gene.

Various methods can be used to assess the presence of inhibitors within biological samples. The efficiency of the PCR in a test sample can be assessed by serial dilution of the sample,[25] although this is practically impossible for every sample of a high-throughput study or when using very small amounts of precious RNA. Alternatively, there are various algorithms that provide an estimate of PCR efficiency from analysis of amplification curves.[26–28] Internal amplification controls (IAC) that co-purify and co-amplify with the target nucleic acid can be used to detect inhibitors as well as indicate template loss during processing.[29] Another approach utilises a whole bacterial genome to detect inhibition from clinical samples.[30]

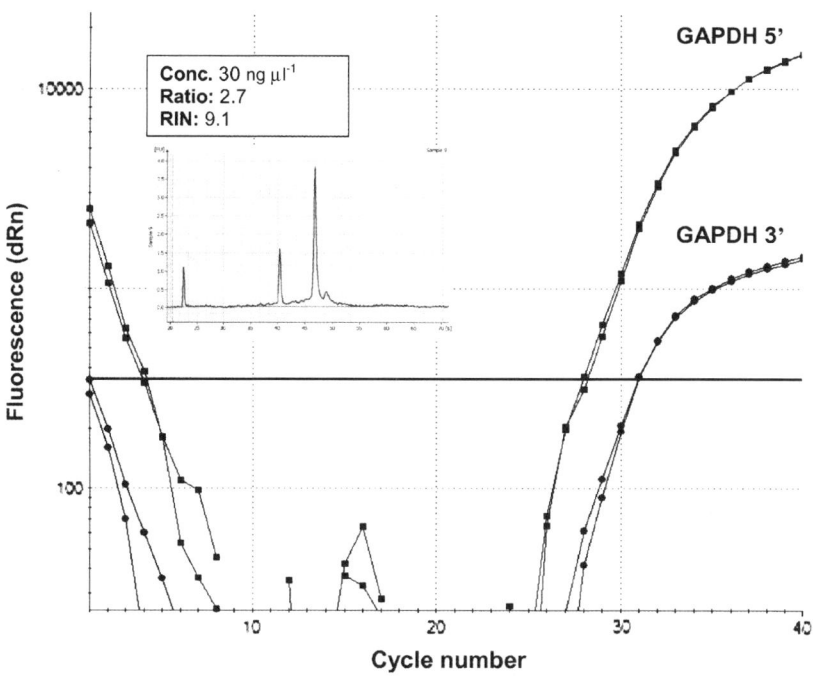

Figure 9.3 Illustration of the effect of inhibitors. EDTA was added to an RNA sample to a final concentration of 125 mM and this was assessed using the GAPDH 3':5' assay for detection of degraded RNA (see Figure 9.2). A clear shift to higher Ct was observed for both the 3' and 5' GAPDH assays with a more pronounced shift of the 5' reaction.

Nolan et al.[31] describe the use of a universal qPCR reference assay, known as SPUD, to identify inhibitors of the reverse transcription or PCR steps by recording the relative Cts characteristic of a defined number of copies of a sense-strand amplicon. An artificial amplicon (SPUD-A) is amplified using two primers (SPUD-F) and (SPUD-R) and the products are detected using a TaqMan™ probe (SPUD-P) (Figure 9.4A). In the presence of water, a Ct is recorded that is characteristic of an uninhibited reaction (dependent on amplicon copies used and technical variabilities). Alongside this reaction, which contains only the SPUD amplicon, reactions are run which contain exactly the same components (SPUD-A, SPUD primers and SPUD probe) together with the unknown test sample (RNA or DNA). Potential inhibitors in the test sample will result in a shift to higher Ct for these reactions when compared to those where the test sample is absent.

Conventional A_{260}/A_{280} measurement does not detect the presence of high concentrations of EDTA that are clearly detrimental to qPCR amplification in an assay-specific manner (data not shown). Interestingly, electrophoretic traces of these samples result in a comparatively low estimate of RNA concentration in the presence of EDTA although these were derived from the

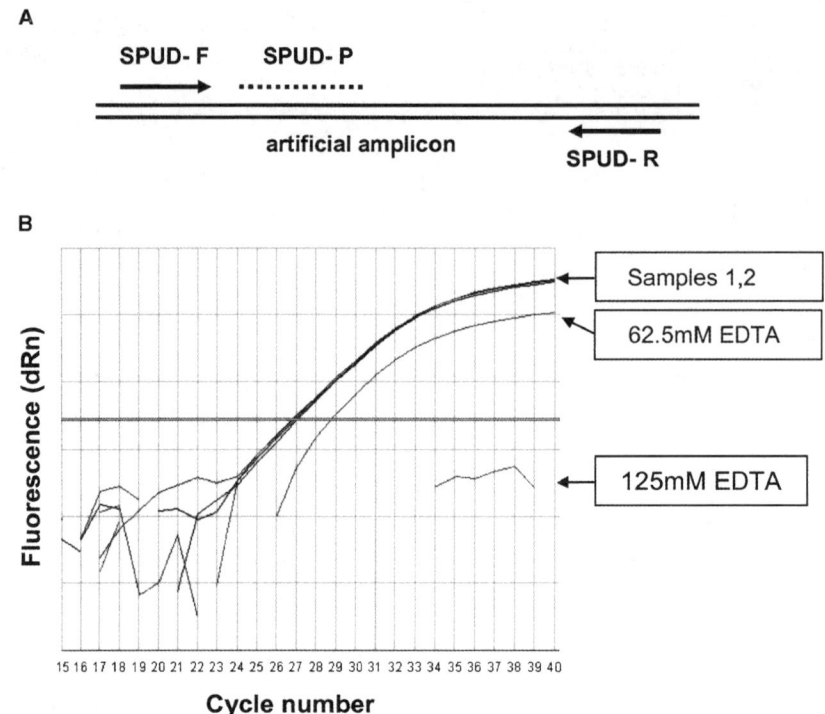

Figure 9.4 Use of the SPUD universal qPCR reference assay. (A) In the presence of water, a Ct is recorded that is characteristic of an uninhibited SPUD reaction. (B) Amplification of the SPUD amplicon in the presence of the intact and degraded samples (samples 1 and 2 respectively) does not affect the assay whereas EDTA at 62.5 mM caused a 2 Ct shift and EDTA at the higher concentration prevented all amplification in the SPUD assay.

purified samples and known to be of equal concentration (110 ng μl^{-1}). It is worthy of note that EDTA suppressed the fluorescence reading by both the Agilent 2100 Bioanalyzer and the Bio-Rad Experion systems. RNA samples containing EDTA at final concentrations of 125 mM and 62.5 mM were included in the SPUD assay alongside purified RNA and purified, degraded RNA samples (Figure 9.4B). Amplification of the SPUD amplicon in the presence of the purified RNA samples and the degraded sample resulted in exactly the same Ct (27). In the presence of EDTA at the lower concentration the Ct is shifted by 2 Cts and at the higher concentration no amplification is detected. In this example the EDTA is sufficient to prevent any amplification of the SPUD amplicon. This demonstrated that the SPUD assay is a useful tool for identification of inhibitors in samples and the application of the technique to screening samples extracted from FFPE sections is shown by Nolan et al.[31] Novak and Huggett have also demonstrated that the system can be used to identify false negative results due to inhibition of the test PCR.[32]

9.4 RNA Quantification

9.4.1 Significance of Quantification

Many downstream molecular biology assays that use RNA are sensitive to template concentration. This can be demonstrated by reverse transcription of a serial dilution of RNA and quantification of specific cDNA targets. An example of this phenomenon is illustrated in Figure 9.5. In this experiment, cDNA was produced from a five-fold serial dilution of total RNA using StrataScript® reverse transcriptase (Stratagene, UK) and random nonamer primers. The quantity of a number of target genes was determined and it is clear that there was not a linear relationship between the initial concentration of RNA and the level of the specific cDNA yield. At the most extreme there is an inverse relationship between the most concentrated RNA sample and the cDNA yield (the first two amplification plots are 'reversed') indicating that high concentrations could inhibit reverse transcription. In a further study RNA was diluted 100-fold, cDNA produced as described previously and the quantity of βactin was determined (Figure 9.6A). The cDNA synthesis was replicated using the same RNA dilution series and duplicate qPCR reactions were run from each independent RT series. As before there was not a linear relationship between the initial RNA and cDNA yield but the yield of βactin was reproducible. A constant number (10^4) of copies of a specific target sequence was added to each RNA dilution sample and cDNA made from the mixture. The number of copies of the spike sequence in each sample was then determined

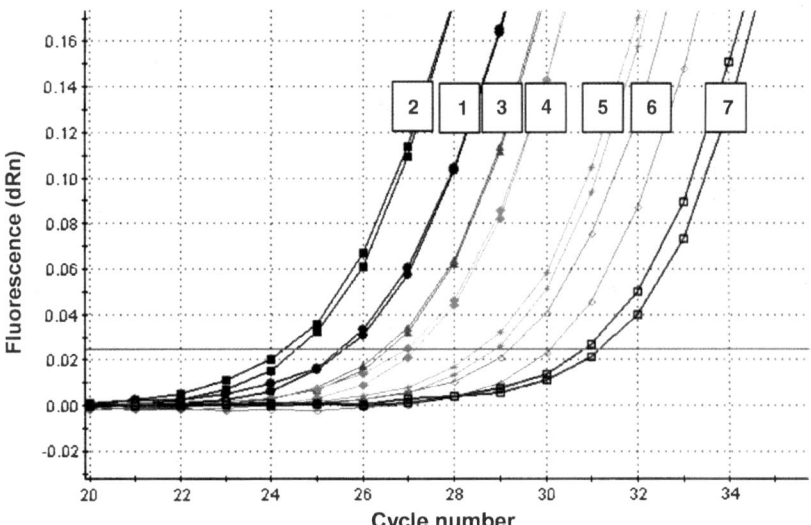

Figure 9.5 qRT-PCR quantification of βactin from cDNA produced from five-fold serial dilution (1–7) of total RNA. The lowest Ct was recorded from the sample containing the second highest concentration of RNA demonstrating that high concentration of RNA could inhibit reverse transcription.

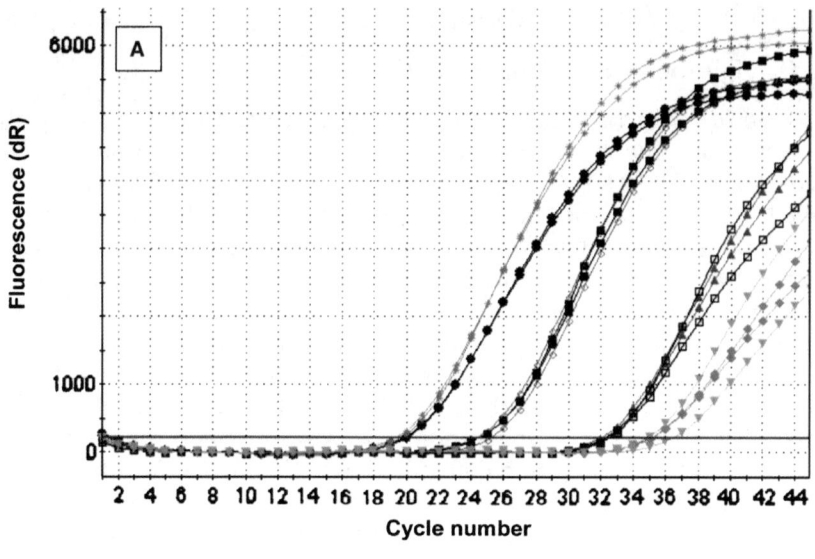
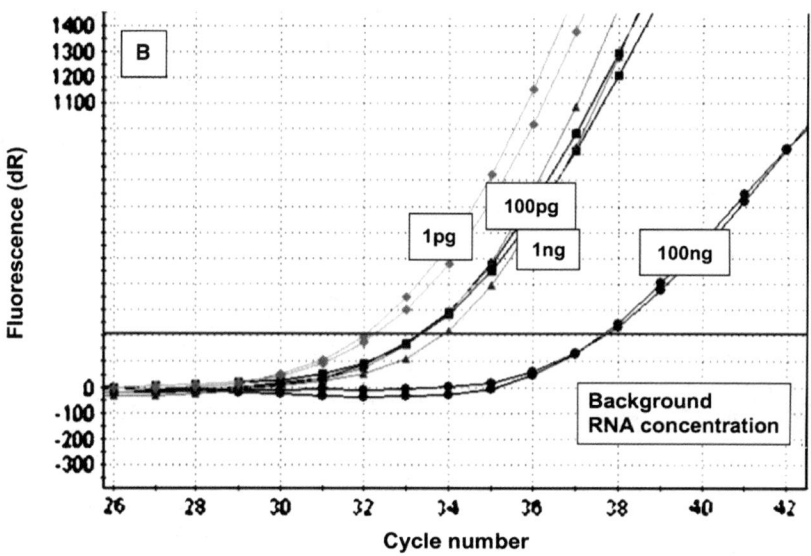

Figure 9.6 Effect of RNA concentration. (A) cDNA was produced from a 100-fold serial dilution of RNA and the quantity of βactin was determined. The cDNA synthesis was replicated using identical conditions. In each case there was not a linear relationship between the initial RNA and cDNA yield. (B) A constant copy number (10^4) of a specific sequence is added to the 100-fold RNA dilution and the specific target detected by qPCR. A higher concentration of spike sequence molecules was detected in the sample containing the lowest concentration of background RNA and a lower concentration of spike molecules was detected in the samples containing a higher concentration of RNA. The same number of copies of the spiked sequence was detected in samples containing both 100 pg and 1 ng.

using qPCR. Since exactly the same number of molecules was added to each RNA sample, the same Ct should be produced. In contrast, more spike sequence molecules were detected in the sample containing the lowest concentration of background RNA and a lower concentration of spike molecules was detected in the samples containing a higher concentration of RNA. The same number of copies of the spiked sequence was detected in samples containing both 100 pg and 1 ng. As observed for the RT reaction, there was a non-linear inverse relationship between the number of spike targets detected and the background concentration of cDNA (Figure 9.6B).

A similar phenomenon was also reported by Stahlberg et al.[33] who also demonstrated that this effect can be relieved by the addition of carriers such as PEG.

9.4.2 Methods of Quantification

It is for these reasons that RNA samples should be quantified after extraction whenever possible. In the absence of a perfect nucleic acid quantification system, the approach which is most suitable for the laboratory should be used. The NanoDrop® system (NanoDrop Technologies, USA) has a wide, dynamic range of quantification but is labour intensive because it only processes a single sample at a time; the chip analysis systems from Bio-Rad and Agilent process up to twelve samples simultaneously and provide measures of RNA quality (but see Section 9.3.1 and below for an assessment of these measurements). The disadvantage of these systems is that they are relatively expensive. When more samples are to be quantified, RiboGreen® staining (Molecular Probes, Invitrogen, USA) is a practical approach. This is a simple binding dye assay and can be carried out using a fluorescence plate reader or any qPCR system that has an integral sample florescence read function.

It has been demonstrated that when exactly the same samples are quantified using different quantification methods the results are wildly different.[34] An example of a comparison between quantification values is shown in Figure 9.7. There are similarities between the NanoDrop® and spectrophotometric quantities because these both use A_{260} conversions. The Agilent Bioanalyzer and Bio-Rad Experion are also similar, although the absolute values derived from the Experion are consistently lower than those from the Bioanalyzer. The RiboGreen® determination of RNA quantities was generally higher than that of any other system. It is striking that the different approaches resulted in a different quantification and that the relationship between these values is not consistent. Whichever quantification system is selected it is critically important to use the same system for all samples which are to be included in a given experiment.

9.5 Effect of RT Experimental Design on qPCR Data

It has been demonstrated that it is necessary to correct for the lack of linearity between the concentration of RNA and the cDNA yield. This can be achieved by addition of carrier[35] or inclusion of exactly the same concentration of RNA

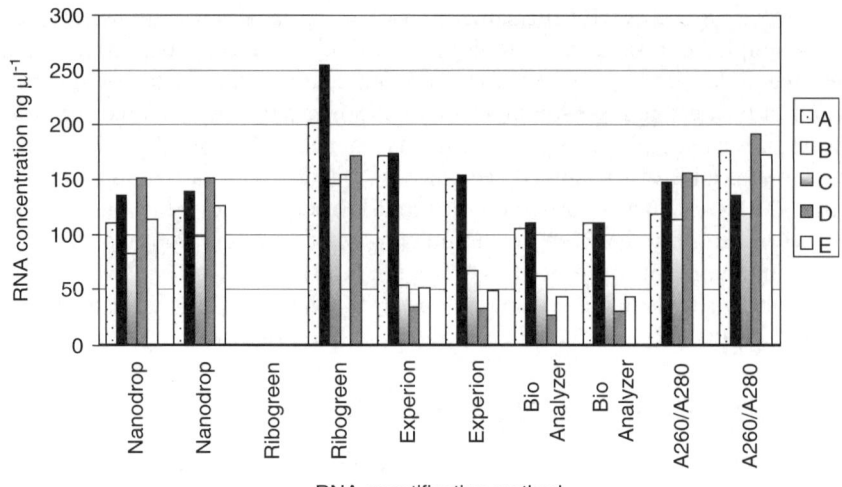

Figure 9.7 Comparability of RNA quantification methods. RNA samples A and B consisted of purified RNA samples, normalised to 100 ng μl^{-1} with reference to Nanodrop® quantification, C was prepared by degradation of sample A, samples D and E were prepared by addition of EDTA to sample A to final concentrations of 125 mM and 62.5 mM respectively. The concentration of each sample was determined from independent duplicate measurements using the Agilent Bioanalyzer, Bio-Rad Experion, Nano-Drop®, RiboGreen® and conventional spectrophotometry reading at A_{260}. The duplicate readings using identical systems are highly reproducible. (Data kindly provided by students attending the qPCR training course at EMBL, Heidelberg.)

into the qRT-PCR reaction. The latter is the most usual and it is reasonable to expect that replicated samples would produce exactly the same cDNA profile. Many experimental designs rely upon periodic acquisition of clinical samples that are processed in batches. The data from a typical experiment are shown in Figure 9.8. Total RNA was extracted from clinical samples, quantified and then global cDNA produced using random nonamer primers. Each batch of clinical samples was processed alongside a positive control RNA sample (human reference RNA; Stratagene). The cDNA samples were then interrogated for the quantities of the transcripts of interest with reference to a serial dilution of human reference RNA (Stratagene). Transcript quantities of the test gene in the cDNA of two batches of clinical samples and both positive controls are shown in Figure 9.8A and the quantification of βactin in the same samples is shown in Figure 9.8B. Analysis of the test gene transcript quantities indicates that this is present at lower levels in the first set of samples than in the second set. The comparison of the quantities in the controls (calibrator reference samples) associated with the two batches reveals that the reverse transcription was less efficient for the first calibrator reference sample. The most common procedure to correct for differences in reverse transcription efficiency is to refer the gene of interest to one or more stable reference genes. This technique is performed with the expectation that the reverse transcription of all transcripts is equally efficient in all samples. Analysis of the quantity of βactin in the same cDNA samples reveals that this is not a safe assumption. In contrast to the test gene profile, a higher yield of βactin is detected in the first batch of samples and this is also reflected in the calibrator reference samples. This phenomenon was explored further by comparison of the quantity of three genes in the identical calibrator reference samples processed on four independent occasions. Despite all practical variables being constant, the relative quantity of different transcripts varied between the apparently identical reverse transcription reactions (Figure 9.8). It is evident from these data that reverse transcription reactions are not always reproducible between batches and that the variability does not maintain the proportionality of transcript quantity.[36–37]

An alternative method for construction of a calibration curve is to dilute total RNA and detect the specific target using target-specific priming. Under these experimental conditions, in contrast to the use of random priming, there

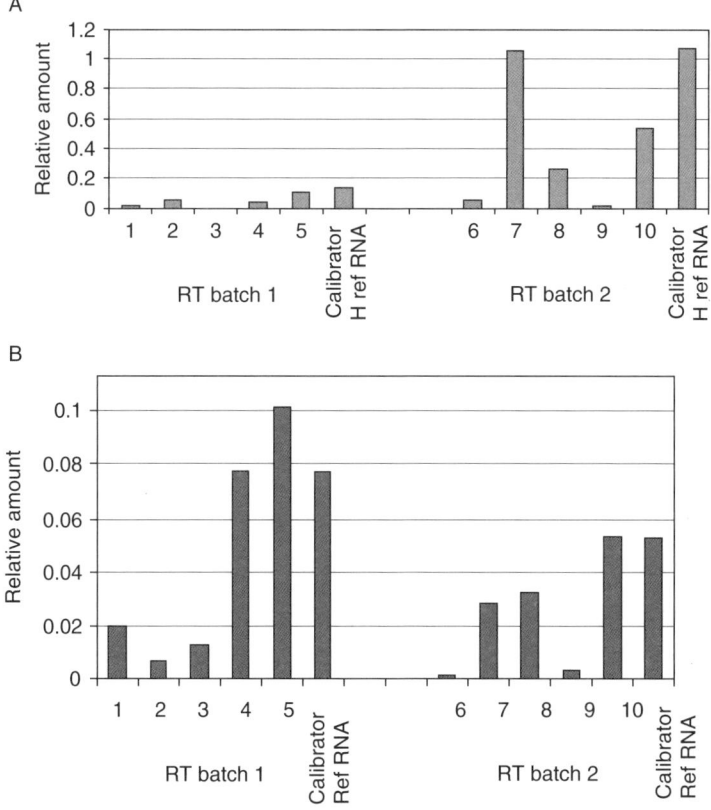

Figure 9.8 *(Continued)*

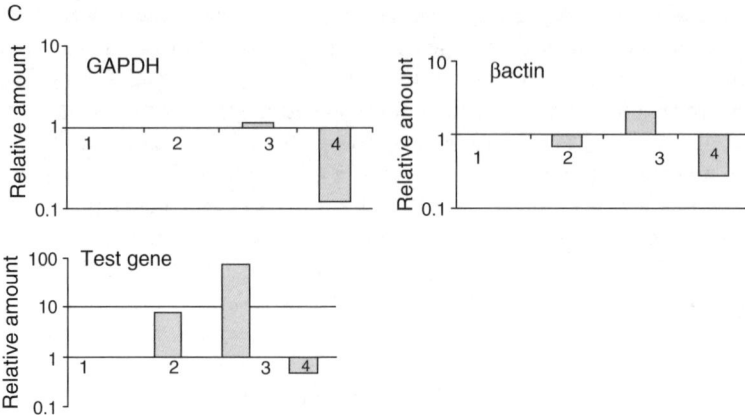

Figure 9.8 The quantity of each specific gene target was determined in each cDNA sample and is expressed relative to the quantity detected in the universal standard curve (constructed from a serial dilution of cDNA). (A) The transcript quantities of the test gene in the cDNA clinical samples and positive controls were different between two batches of RT reactions such that the test gene transcript quantities are lower in the first set of samples than in the second set. (B) The quantification of βactin in the same samples indicates that a higher yield of βactin is detected in the first batch of samples and this is also reflected in the calibrator reference samples. (C) The quantity of three genes, GAPDH, βactin and test gene, were determined in identical calibrator reference samples processed on four independent occasions. The graphs show the data from each quantification relative to the quantity of the target recorded in the first experiment. There were similar quantities of GAPDH in the first three replicated cDNA samples with a relative decrease of around eight-fold in the 4th sample. There was more variability in the βactin quantities in samples 2 and 3, with only a six-fold decrease recorded in the 4th sample. There was much more variability in the quantities recorded for the test gene with a 72-fold increase in the 3rd sample and a three-fold decrease recorded in the 4th sample. The RT reaction was variable between batches and transcript proportionality was not maintained.

is a linear relationship between the initial concentration of RNA and the yield of specific target (Figure 9.9).

9.6 Conclusion

The qRT-PCR is undoubtedly the method of choice for quantification of specific RNA targets. However, in order to produce reliable mRNA quantification data it is critical to ensure that each stage of the process is optimal; all processes require validation including RNA extraction and quantification, template quality assurance assessment, reverse transcription reproducibility and finally qPCR assay optimisation. Until each of these processes is standardised and the information to demonstrate that these procedures have been carefully

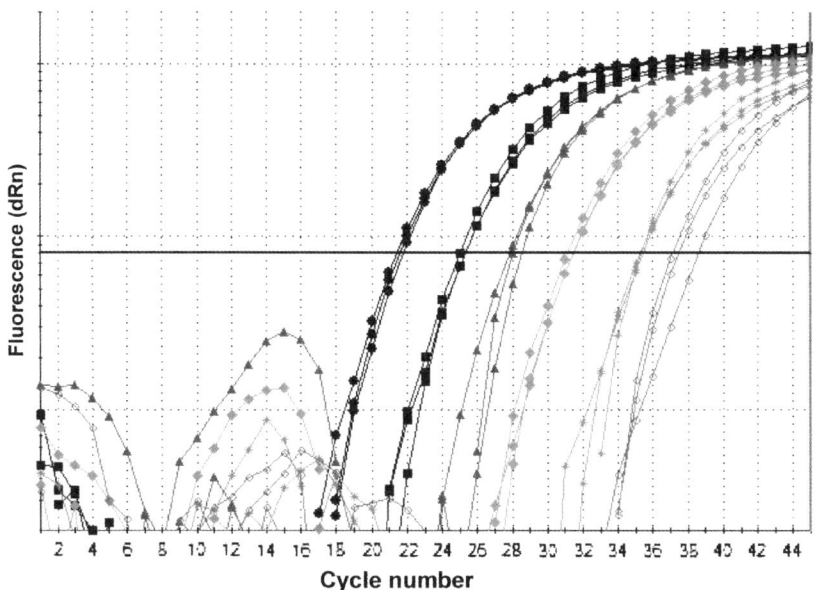

Figure 9.9 A calibration curve constructed by dilution of total RNA and GAPDH transcript detected using target specific priming. In this case there is a linear relationship between the initial concentration of RNA and the yield of the specific target.

controlled is included in peer-reviewed papers it will remain almost impossible to compare the wide range of reports due to lack of technical compatibility. Worse still, lack of control over any one of the required procedures can lead to meaningless numbers gaining apparent validity due to statistical analysis. Simply, validation matters.

References

1. G. L. Mutter, Z. Zahrieh, C. Liu, D. Neuberg, D. Finkelstein, H. E. Baker and J. A. Warrington, *BMC Genomics*, 2004, **5**, 88.
2. E. S. Lader, 2001, US patent 6,204,375.
3. F. Lewis, N. J. Maughan, V. Smith, K. Hillan and P. Quirke, *J. Pathol.*, 2001, **195**, 66.
4. Y. N. Park, K. Abe, H. Li, T. Hsuih, S. N. Thung and D. Y. Zhang, *Am. J. Pathol.*, 1996, **149**, 1485.
5. D. Bresters, M. E. Schipper, H. W. Reesink, B. D. Boeser-Nunnink and H. T. Cuypers, *J. Virol. Meth.*, 1994, **48**, 267.
6. S. A. Bustin, unpublished observations.
7. T. E. Godfrey, S.-H. Kim, M. Chavira, D. W. Ruff, R. S. Warren, J. W. Grey and R. H. Jensen, *J. Mol. Diag.*, 2000, **2**, 84.

8. D. von Smolinski, I. Leverkoehne, G. von Samson-Himmelstjerne and A. D. Gruber, *Histochem. and Cell Biol.*, 2005, **124**, 177.
9. M. Macabeo-Ong, D. G. Ginzinger, N. Dekker, A. McMillan, J. A. Regezi, D. T. Wong and R. C. Jordan, *Mod. Pathol.*, 2001, **15**, 979.
10. C. Williams, F. Pontén, C. Moberg, P. Soderkvist, M. Ulén, J. Pontén, G. Sitbon and J. Lundeberg, *Am. J. Pathol.*, 1999, **155**, 1467.
11. J. Davis, B. Maqsodi, W. Yang, Y. Ma, Y. Luo and G. McMaster, *J. Biomol. Tech.*, 2007, **18**, 9.
12. W. Yang, B. Maqsodi, Y. Ma, S. Sui, K. L. Crawford, G. McMaster, F. Witney and Y. Luo, *Biotechniques*, 2006, **40**, 481.
13. A. Schroeder, O. Mueller, S. Stocker, R. Salowsky, M. Leiber, M. Gassman, S. Lightfoot, W. Menzel, M. Granzow and T. Ragg, *BMC Mol. Biol.*, 2006, **7**, 3.
14. S. Fleige and M. W. Pfaffl, *Mol. Aspects Med.*, 2006, **27**, 126.
15. C. A. Pérez-Novo, C. Claeys, F. Speleman, P. Van Cauwenberge, C. Bachert and J. Vandesompele, *Biotechniques*, 2005, **39**, 52.
16. T. Nolan, R. E. Hands and S. A. Bustin, *Nat. Prot.*, 2006, **1**, 1559.
17. H. Auer, S. Lyianarachchi, D. Newsom, M. I. Klisovic, G. Marcucci and K. Kornacker, *Nat. Genet.*, 2003, **35**, 292.
18. P. Radstrom, R. Knutsson, P. Wolffs, M. Lovenklev and C. Lofstrom, *Mol. Biotechnol.*, 2004, **26**, 133.
19. J. Lefevre, C. Hankins, K. Pourreaux, H. Voyer and F. Coutlee, *J. Med. Virol.*, 2004, **72**, 132.
20. E. Sunen, N. Casas, B. Moreno and C. Zigorraga, *Int. J. Food Microbiol.*, 2004, **91**, 147.
21. I. R. Perch-Nielsen, D. D. Bang, C. R. Poulsen, J. El-Ali and A. Wolff, *Lab Chip.*, 2003, **3**, 212.
22. J. Jiang, K. A. Alderisio, A. Singh and L. Xiao, *Appl. Environ. Microbiol.*, 2005, **71**, 1135.
23. R. A. Guy, P. Payment, U. J. Krull and P. A. Horgen, *Appl. Environ. Microbiol.*, 2003, **69**, 5178.
24. A. Stahlberg, N. Zoric, P. Aman and M. Kubista, *Expert Rev. Mol. Diagn.*, 2005, **5**, 221.
25. A. Stahlberg, P. Aman, B. Ridell, P. Mostad and M. Kubista, *Clin. Chem., (Washington, DC, US)*, 2003, **49**, 51.
26. A. Tichopad, M. Dilger, G. Schwarz and M. W. Pfaffl, *Nucleic Acids Res.*, 2003, **31**, e122.
27. C. Ramakers, J. M. Ruijter, R. H. Deprez and A. F. Moorman, *Neurosci. Lett.*, 2003, **39**, 62.
28. W. Liu and D. A. Saint, *Biochem. Biophys. Res. Commun.*, 2002, **294**, 347.
29. B. L. Pasloske, C. R. Walkerpeach, R. D. Obermoeller, M. Winkler and D. B. DuBois, *J. Clin. Microbiol.*, 1998, **36**, 3590.
30. J. L. Cloud, W. C. Hymas, A. Turlak, A. Croft, U. Reischl, J. A. Daly and K. C. Carroll, *Diagn. Microbiol. Infect. Dis.*, 2003, **46**, 189.
31. T. Nolan, R. E. Hands, B. W. Ogunkolade and S. A. Bustin, *Anal. Biochem.*, 2006, **351**, 308.

32. T. Novak, R. F. Miller, A. Pooran, M. Hoelscher, M. Gerhardt, F. Minja, A. Zumla and J. Huggett, 2007, Poster presentation, Proceedings 3rd International qPCR Symposium, Freising, Germany.
33. A. Stahlberg, J. Hakansson, X. Xian, H. Semb and M. Kubista, *Clin. Chem., (Washington, DC, US)*, 2004, **50**, 509.
34. S. A. Bustin and T. Nolan, *J. Biomol. Tech.*, 2004, **15**, 155.
35. A. Stahlberg, M. Kubista and M. Pfaffl, *Clin. Chem., (Washington, DC, US)*, 2004, **50**, 1678.
36. T. Nolan and H. A. Lacey, unpublished observations.
37. H. A. Lacey, T. Nolan, S. L. Greenwood, J. D. Glazier and C. P. Sibley, *Placenta*, 2005, **26**, 93.

CHAPTER 10
Microarrays

SALLY L. HOPKINS AND CHARLOTTE L. BAILEY

LGC, Queens Road, Teddington, TW11 0LY

10.1 Introduction

The use of microarrays combines the specificity achievable with classical hybridisation techniques[1,2] with the throughput potential of massively parallel arrays from the semiconductor industry, to create a system for looking at thousands of genes in a single experiment. This was the beginning of biochip technology that has grasped the imaginations of scientists in a multitude of disciplines across the globe and has expanded exponentially in the decade following the early developments (see Figure 10.1). Microarray technology allows an unparalleled view of biological systems by allowing characteristics of whole genomes, such as gene expression levels, SNP status and methylation patterns, to be examined in just a few days, compared to the weeks it used to take to examine a single gene using traditional techniques. Once the data has been acquired, however, it can be complex to analyse, necessitating the coming together of many scientific disciplines, including biologists, statisticians and computer scientists, in order to produce sound, biologically meaningful conclusions.

This initial section gives a brief introduction to microarrays and their applications. The remainder of the chapter will then concentrate on the difficulties the scientific community has in fulfilling the full potential of microarray technology along with standardisation initiatives and quality controls being developed to help them. The main focus of the chapter is traditional two-colour glass slide DNA microarrays, especially those determining gene expression. The chapter is intended as an introductory guide to microarrays and standardisation, providing information on current issues and activities to improve measurement confidence in the technology.

10.1.1 What are Microarrays?

Microarrays are based on the same principles as traditional Southern blot assays (however, using the reverse format) where a sequence of DNA is used to

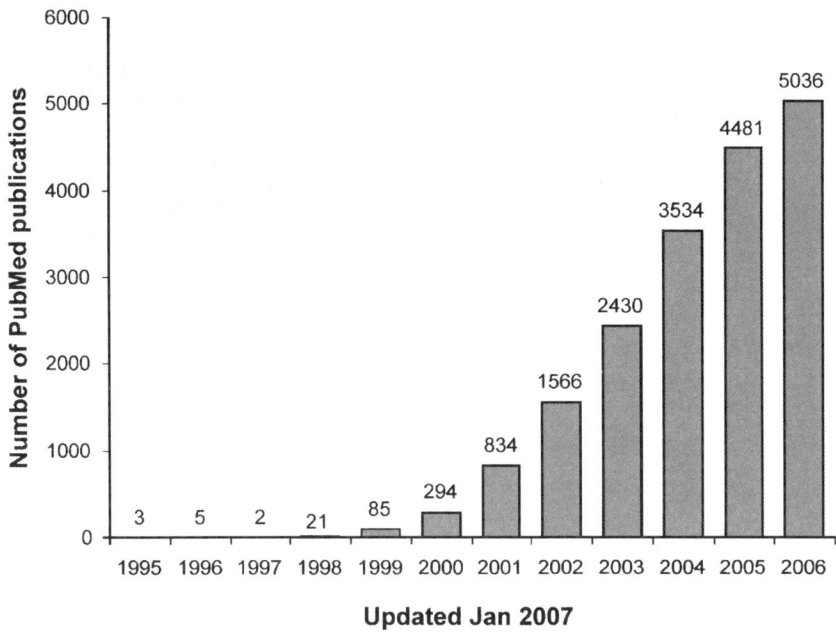

Figure 10.1 Number of microarray publications listed in PubMed each year from 1995 to 2006.

search for the presence of a complementary sequence in a mixed DNA population. Unlike Southern blots, which use flexible membranes as a support for the assay, microarrays utilise solid supports such as glass slides, beads or even bar-coded micro particles and have progressed from using only DNA probes to a variety of other biological materials, such as proteins and tissue.[3] Microarrays have become the method of choice so quickly in many biological fields due to the number of probes it is possible to place and analyse on a single microarray compared to traditional techniques. Thousands of DNA probes can be placed on a single glass slide, for example, and can therefore be analysed within a single experiment (Figure 10.2).

Bead and micro-particle-based arrays are solution-based array systems and will not be the focus of this chapter.[4–6] Solid phase array systems are one of the most common array types being used today and contain multiple biological probes spotted onto a surface in a highly regular and specific order, such that the probe contained in each spot is traceable.

The sample(s) of interest are labelled, usually with a fluorescent dye. A hybridisation then takes place, where the labelled fragments from the target mixture should bind to their immobilised complementary probe. Two or more samples can be hybridised at the same time depending on the type of array, and by measuring the different fluorescent intensities associated with each spot on the array, the relative abundance of specific sequences within the samples can be determined. The fluorescent signals from the hybridised targets are read

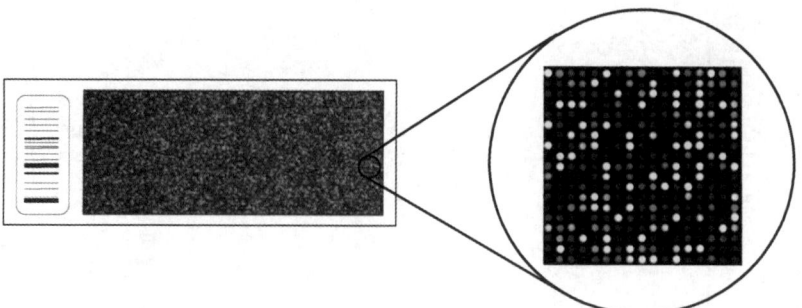

Figure 10.2 Illustration of a high-density glass slide array. A single glass slide array can contain tens of thousands of DNA probes, each seen as a spot on the microarray. The slide shown here has approximately 20 000 spots.

using a microarray scanner and translated into a computer image of the microarray. Specialised computer software is then necessary to extract primary data from these images, normalise the data to correct for the influence of experimental variation and then mine and model the data in order to gain meaningful biological conclusions. Figure 10.3 shows a schematic of the microarray process.

10.1.2 A Note about Nomenclature

There are currently two conventions of nomenclature within the microarray community regarding the definition of the 'probe' and the 'target'.

In one nomenclature system, the *probe* is used to refer to the labelled nucleic acid from each sample, while the *target* refers to the immobile DNA on the array surface. This directly translates from the physical placement of the target and probe in a traditional Southern blot terminology; the *target* being fixed to the membrane and the *probe* being the unbound component.

More recently, however, a second system has come into use, emulating the function of both target and probe in a traditional blot. This defines the *probe* as the known DNA sequence tethered to the array surface, and the *target* as the more complex, free labelled nucleic acid.

When reading publications or reconstructing experiments, it is vital to know which nomenclature system the author is using in order to correctly interpret the data or protocol. The second system of nomenclature will be adopted throughout this chapter.

10.1.3 Types of Microarrays

There are currently many different types of microarrays being utilised in the scientific research community, using many different target materials, such as DNA, RNA, proteins,[7,8] small molecules and tissues. DNA microarrays have been the most widely used over the last ten years, and may utilise a variety of

Microarrays

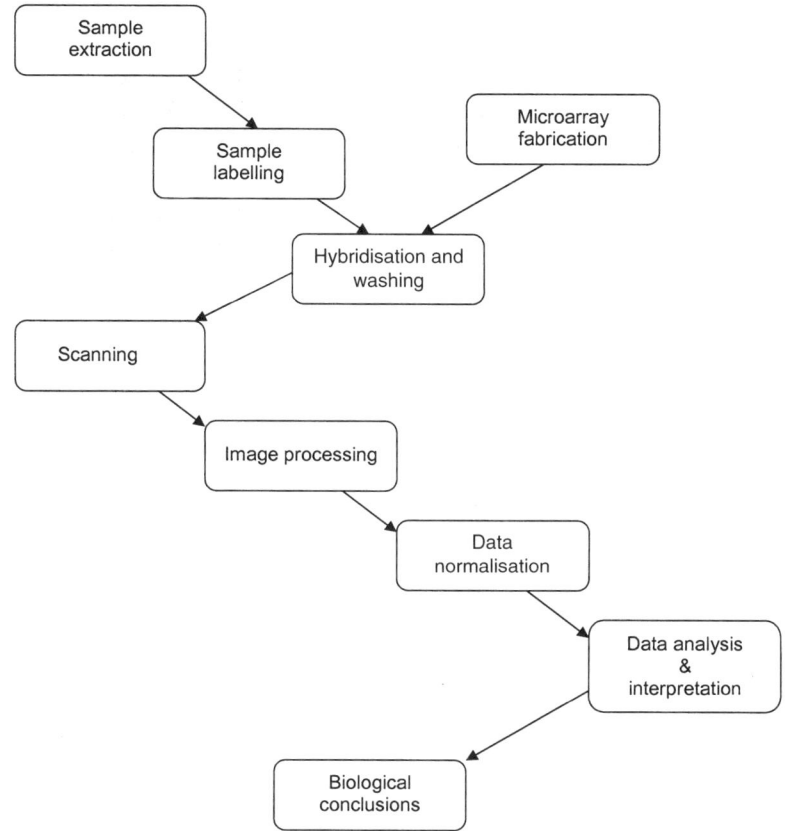

Figure 10.3 Schematic diagram showing the main stages of the microarray analysis process.

probe types including cDNA and oligonucleotide sequences. The format of the array may be either solid phase, where the probes are arranged on a solid surface, or liquid phase array, where beads or similar particles are use to tether probes, and the hybridisation takes place in solution. A variety of substrates may be employed in construction of solid phase arrays, including glass slides, microtitre plate wells and wafers, such as the high-density oligonucleotide arrays that are commercially developed by Affymetrix.

The probes for cDNA microarrays consist of PCR products, derived from cDNA clones.[9] These are then spotted onto glass slides. This spotting process can be achieved by either physical contact deposition or non-contact printing methods. Solid phase arrays are commonly manufactured using microlithography, a light-chemical synthesis process. Multiple probes are synthesised onto a large sheet, which is then split up into individual arrays and packaged into plastic cartridges.

Glass slide arrays are currently the most popular format for scientists spotting their own arrays, due to their lower cost, compatibility with many fluorescent scanning systems, large surface spotting area and flexibility of content, although they are more susceptible to scratching and small particle contamination throughout the experimental process. The solid phase standardised high-density arrays are comparatively higher cost and are limited to a custom scanning format and array content. They do, however, allow a higher level of consistency throughout the experimental process due to constant reagent volumes and are less prone to damage because of protective casings. The choice of array is very much dependent on the experiments and answers being sought by the scientists and should be given a high level of consideration before either setting up a facility in-house, or buying commercially prepared arrays.

10.1.3.1 *Applications of DNA Microarrays*

The major applications of DNA microarrays fall into four categories:[10]

a) Gene expression profiling

These microarrays use RNA from a complex sample, such as body fluids or bacterial isolates, to reveal the expression patterns of thousands of genes. This gives rise to an expression profile or signature for that particular target sample. Figure 10.4 shows some of the possible outputs of gene expression microarrays.

b) Genotyping

Genomic DNA, amplified by PCR, is used as the target. The genotype for hundreds or thousands of genetic markers can be determined in a single hybridisation.

c) DNA sequencing

Genomic DNA is again amplified and then applied to 're-sequencing' microarrays. Thousands of base pairs of DNA can be screened for mutations in specific genes whose normal sequence is already known.

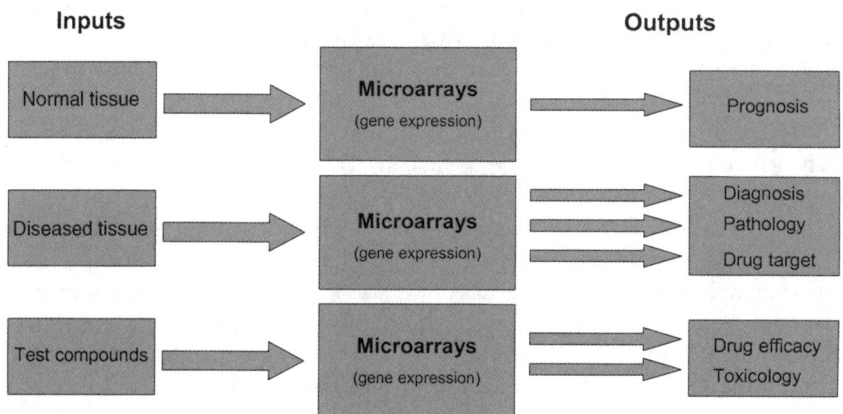

Figure 10.4 Diagram to illustrate some typical inputs and outputs of gene expression microarrays.

d) CGH arrays

Comparative genomic hybridisation (CGH) is a molecular method for the analysis of copy number changes (gains/losses) in the DNA content of cells. It is one of the first real applications for arrays outside the research setting.

The method is based on the hybridisation of fluorescently labelled test DNA and normal/control DNA to a microarray consisting of characterised genomic probes.

The probes on the array represent specific genomic loci and are selected and orientated in such a way as to produce high resolution coverage of the genome under investigation. This technology has most commonly been used to detect chromosomal amplifications and deletions in cancer.

10.1.3.2 Impact of Applications

The greatest impact to date has been using DNA microarrays in basic biological gene expression studies,[11] which have helped to facilitate research in many areas of biology. One such field is that of disease classification.[12–14] Gene expression profiling of diseased cells can provide a very detailed view of the class, subclass and the stage of the disease. By classifying patients by their gene expression profiles, clinicians may be able to assign correct treatment, as different disease states may respond best to different treatments. It may also be possible for clinicians to gauge the prognosis of a patient with greater accuracy using the profile. Characterisation of the many mutations in specific genes greatly increases the scope for a precise molecular diagnosis in single gene and more complex genetic diseases.

Microarrays are being used in the area of drug discovery by many companies.[13] Genome-wide expression profiling allows drug developers to elucidate complex disease mechanisms, thus identifying potential new targets for drugs. Toxicogenomics uses microarrays to monitor any changes in gene expression seen in response to particular toxins, such as drug treatment. The expression profiles obtained can then be compared to databases of toxic cell expression profiles, allowing scientists to predict whether a drug will have adverse side effects. This allows the prediction of a drug's toxic side effects far earlier in development than traditional methods, enabling pharmaceutical companies to cancel drug development, if toxicity is indicated, before proceeding to costly clinical trials.

Pharmacogenomics looks at the inherited variations in genes dictating drug response, exploring ways in which these variations can predict the patient's response to the drug and is another field which has evolved greatly since the introduction of microarray analysis.[11] Knowledge of a patient's drug metabolising gene variants can help determine the appropriateness and dosage of drugs prescribed to patients. Microarrays are able to provide this genome-wide picture of the patient's genetic makeup. There are also disease-specific microarrays already on the market, such as Amplichip CYP450, which looks at variation in two genes that play a role in the metabolism of many widely prescribed drugs (Roche Diagnostics, USA)[15,16] and the MammaPrint® array

(Agendia, Netherlands) testing for breast cancer prognosis,[17] both of which have obtained US Food and Drug Administration (FDA) approval.

Analysis of genomic DNA has also led to detection of differential copy numbers of genes between samples, mapping of DNA binding sites, methylation status of genes, detection and monitoring of microRNA and small interfering RNA, as well as the detailed characterisation of microbial pathogens. These studies could have big impacts on the diagnosis, tracing and monitoring of infectious diseases as well as environmental monitoring.[11]

10.2 Technology Status

Microarray technology has provided the scientific community with a very powerful tool. Using microarray analysis facilitates the global analysis of gene expression to identify trends and possible interactions and the past decade has seen microarrays evolve to become a standard method of choice for many genomic applications. The high-throughput capability of microarrays significantly reduces time and cost spent running assays compared to using conventional techniques, such as northern blotting. However, obtaining consistent and high-quality data from microarray experiments from which valid biological conclusions can be drawn presents a significant technical challenge and the exploitation of the full potential of the technology continues to be hampered by this. It proves extremely difficult to obtain consistent results because of the complexity of the process and the current lack of standardisation.

10.2.1 Current Problems

The overall measurement process, critical to the reproducibility and comparability of microarray data, may be broken into a number of consecutive stages (Figure 10.5), each of which has many sources of variation which will affect the quality of the biological results and conclusions drawn from an experiment.

- Microarray fabrication – the preparation of a suitable solid phase, the synthesis or preparation of probe molecules and the high-density arraying of probes. Some of the factors which can impact here are humidity, printing solutions, pin heads and probe suspensions.
- Target preparation and labelling – RNA extraction and quality assessment followed by fluorescence-based labelling of target material. Factors to be considered include RNA extraction method, RNA quality, labelling technology, fluorescence yield and uniformity of dye incorporation.
- Probe target hybridisation – hybridisation is affected by a variety of physico-chemical and thermodynamic properties such as reagent composition, temperature and hybridisation kinetics. There is a need to characterise, control and correct for sequence-specific effects;
- Signal production/detection – uncertainties introduced during the target-labelling step are compounded during the signal detection procedure, arising from the detection and measurement systems employed;

Microarrays 215

Figure 10.5 Flow diagram of the microarray process, including QC considerations.

- Image processing, normalisation and interpretation – these are the last major stages in the generation of microarray results; all procedural errors are compounded by uncertainties in the normalisation step.

Sample quality, labelling protocol, hybridisation conditions, scanning instrument, image acquisition and processing, data normalisation and analysis, quality assessment of data and interpretation of results all contribute to the overall uncertainty of the conclusions drawn from microarray experiments.

The rapid uptake of this technology by the scientific community has meant that development of appropriate standards has lagged far behind that of the

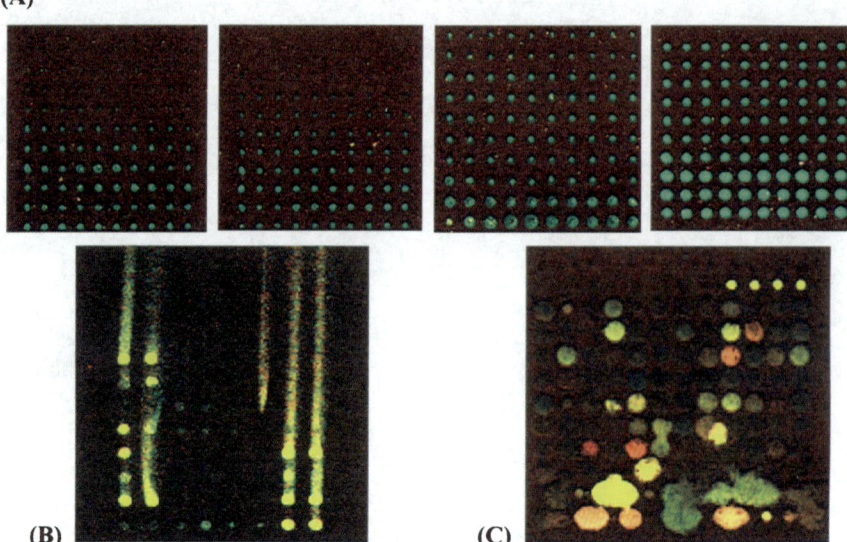

Figure 10.6 Problems that can occur during the microarray fabrication. (A) The four blocks shown above are from a single glass slide spotted with fluorescent dye. All the blocks are printed using a different pin within a single pin head, each block is printed using a single aliquot of dye and each pin prints a group of 10 × 10 spots with a single charge of dye. Each block is supposed to contain an identical 10 × 10 pattern with the same amount of dye in each spot. Variations seen in spot morphology and amount of dye deposition may result from problems with the humidity of the environment or overloading of the pin heads, and have the potential to significantly affect the uniformity of results obtained. (B) 'Comet' spots and (C) spreading of spots can be seen when the spotting buffer and the slides are incompatible.

technology itself. Due to the availability of spotting instruments and lower costs, more research groups are able to produce their own microarrays. However, the trend seems to be moving away from this because of awareness over spotting consistency and quality control (see Figure 10.6). In-house fabrication gives research groups the advantage of far greater flexibility with their experiments, as the technology can provide platforms for numerous applications. Nevertheless, comparability is key when it comes to assessment of experimental procedures and results.[18] Comparing results from seemingly identical experiments between different laboratories, operators or even days can prove very challenging, and as there are few standards in production, application or interpretation, and a variety of analytical platforms, this exacerbates the problems of comparability.[19]

10.2.2 Controlling Experimental Uncertainties

This section is by no means an exhaustive list and concentrates mainly on data generation and normalisation rather than downstream data interpretation.

Much of this section is derived from guidelines produced by LGC as part of the Measurements for Biotechnology programme and published on the MfB website as well as other references.[11,20] DNA microarray platforms and applications are also undergoing rapid development and are dependant on the type and purpose of the experiment being performed. It is hoped, however, that some of the issues raised will be especially useful to scientists that are new to this area, and may be helpful as a starting point for carrying out microarray experiments. Some of these problems are clearly evident, whereas others are less apparent.

This section is also intended to give an idea of the many different steps, and possible pitfalls, which can be encountered when performing a microarray experiment.

10.2.2.1 Experimental Design

The first step on the way to obtaining confidence in microarray data is good experimental design.[21] Meticulous planning of every step of the process is critical as it will save both time and financial resources by eliminating the need for unnecessary repetition of experimental steps and subsequent loss of precious sample material and expensive kits and arrays. It is essential to optimise experimental conditions and practice handling on test sets of arrays when using new arrays, techniques or equipment, regardless of the analysts' experience. This ensures the highest quality results are obtained from the experimental arrays being used.

Experimental design in the context of microarray experiments requires consideration of a number of factors, including the allocation of samples to the array, and the amount of replication. There are many drivers to perform replication, one being the potential to provide more reproducible results; however, there is no current consensus regarding the amount of replication needed. Some investigators[22] promote three or more replicate measurements for each sample whereas others[23] feel that two replicate measurements are sufficient. The required level of both technical and biological replication for an experiment should be determined, so that there is sufficient confidence in the information from each particular sample and microarray combination (technical repeatability), and also sufficient numbers of different samples are analysed to avoid any peculiarities or variation in a single sample from biasing the interpretations that are drawn from the results (biological variability). In a resource-limited situation, it may be preferable to analyse a larger number of different samples, rather than to perform many repeats of an individual array and sample combination, for example to avoid individual gene expression variability overly affecting the results, as biological replication is usually most important. To ensure there is some indication of the technical reliability of a microarray analysis, it is also advantageous to have at least duplicate probes spotted for each gene (as discussed in Section 10.2.2.2) and two replicate sample hybridisations.[24]

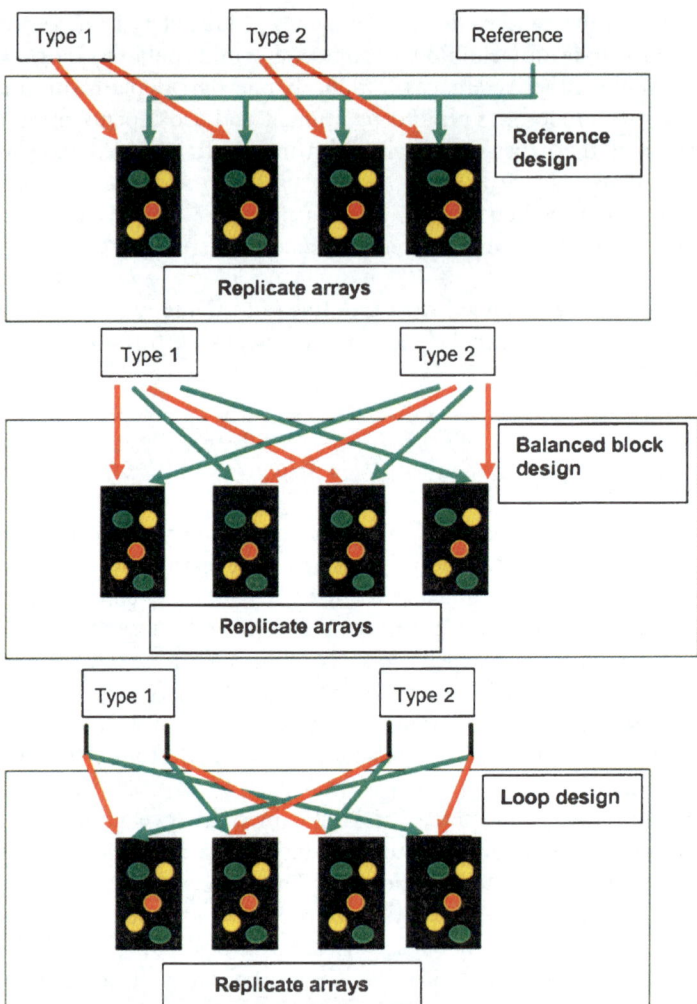

Figure 10.7 Schematic representation of three commonly used experimental designs. A number of considerations influence the approach for comparing expression profiles of two samples, including the available time, resources and amount of sample. In some circumstances it may be preferable to compare samples to a standard reference RNA, rather than directly compare two precious or limited samples. Here the differential labelling of the samples, and the order in which they are applied to replicate arrays, is indicated by the red (Cy5) and green (Cy3) coloured arrows, with independent sampling represented by black lines. Adapted from reference [25].

There are three common types of design that should be considered when using cDNA arrays to identify differentially expressed genes. The three approaches are known as the 'reference design', which is the most widely used design, the 'balanced block design' and the 'loop design' (Figure 10.7). These three approaches will not be discussed in detail in this chapter, however more detail can be found in the following papers.[24,25] The main advantage of using one of these systems is that they have all been developed to provide an unbiased estimate of difference in gene expression levels, although the choice of best design will depend on the amount of time and material available for the investigation.

To avoid any potential differences in array quality that may affect data comparability, it is recommended to use arrays from the same batch during a series of related experiments, such as replicate hybridisations. Differences in probe deposition could result in a gene transcript giving a good signal on an array from one batch, but a poor signal on an array from another batch. If this is the case for a gene of interest assayed by replication on arrays from different batches, valuable information may be lost. Generally, arrays from the same batch should have been subjected to the exact same treatment and conditions. Using arrays from a single batch only, throughout a series of related experiments, eliminates yet another potential source of variability from the experimental procedure. However, true batch-to-batch variability between commercial arrays may significantly bias results, and thus performing checks on comparability of batches is recommended.

10.2.2.2 Microarray Layout and Content

Whether spotting 'in-house' or buying commercially, knowledge about the exact sequence of any probe spotted onto arrays is imperative. However, one should be aware that with cDNA arrays there is no control over the exact sequence of the clone. There may, therefore, be poor specificity as many gene-coding sequences contain regions that may be common with other genes. The sequence of each spot on an oligonucleotide array should be known and this information is usually provided by manufacturers when purchasing commercially. Each probe has a specific affinity for its target based on the GC content of the probe/target duplex, secondary structures and nearest-neighbour interactions for instance. Stronger binding will therefore lead to higher signal intensities. For this reason it is not valid to compare expression levels between genes within a sample or between different probes within the same gene.

In general, the presence of replicate features, often spotted at different co-ordinates across an array, enables an assessment of intra-array variability. Replicates also minimise the impact of distorted signal intensities affecting only parts of the array. Comparison of replicate features may reveal potential outliers, which can subsequently be excluded from data analysis. Combined with replication at array level, this will allow for a comparison of intra- *versus* inter-run variability using statistical tools, such as Analysis of Variance (ANOVA).

It is worth noting that some array manufacturers only spot one feature per gene on an array. For the reasons listed above, data from such arrays are more susceptible to distortion caused by hybridisation-related phenomena than arrays containing replicate features. Such considerations should also be made when designing and producing arrays in-house.

10.2.2.3 Target Quality

RNA quality is a critical factor in microarray-based experiments, and assessing the RNA purity and level of degradation prior to cDNA synthesis is very important, as already described for quantitative reverse transcription PCR (qRT-PCR) analyses in Chapter 9. Inspection of RNA quality is especially important following shipping, as transportation can affect the level of degradation in the sample. The sample purity can be checked by spectrophotometric analysis (OD_{260}/OD_{280} ratio), and RNA integrity by size analysis using gel electrophoresis or using the RNA LabChip with the Agilent 2100 Bioanalyzer (Agilent Technologies, USA). The latter provides a detailed electropherogram showing the size distribution of ribonucleic/nucleic acids within the sample (Figure 10.8).

When analysing total RNA, the areas beneath the 28S and 18S ribosomal peaks are used to calculate the 28S/18S ratio. This value has been used historically as a measure of RNA quality and should be no less than a value of 1 for total RNA used as template for cDNA synthesis. The 28S/18S ratio, however, is not a perfect indicator of RNA quality so it is important to additionally assess the level of degradation by visual inspection of the electropherogram.

The Agilent 2100 Bioanalyzer expert software now offers an RNA Integrity Number (RIN) in the eukaryotic total RNA Nano Assay, which gives a numeric value between 1 and 10 for the integrity of a RNA sample.[26] A RIN of 1 is very degraded and a RIN of 10 indicates intact RNA. The software takes into account ten areas of the electropherogram, including the 18S and 28S peaks, in order to calculate the RIN and provides an objective measure of RNA quality that is of use in quality control. The RIN enables comparison of

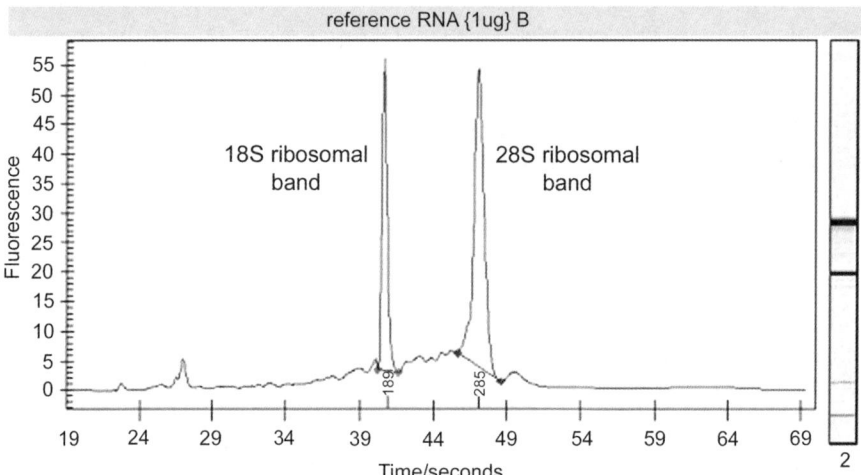

Figure 10.8 Example of an electropherogram to assess RNA integrity. 0.5 μg of total RNA was analysed using an RNA 6000 Nano LabChip and Agilent 2100 Bioanalyzer.

samples, before or after shipment, for example, and comparison from one lab to another, which is more challenging when relying on simple measurement of ribosomal ratios. The RIN is less susceptible to variability across runs on different instruments, and less susceptible to differences in dilutions for the same sample. It is vital, however, that a validation is carried out to correlate RINs with downstream experiments that work or those that do not have the desired outcome, in order to gain a RIN threshold for the particular experiments being carried out in the laboratory. RIN thresholds may vary, for instance between a microarray experiment and qRT-PCR (see Chapter 9). Samples falling below the RIN threshold may then be discarded in order not to waste money on downstream experiments that will fail. Validation of the RIN threshold should be repeated if experimental parameters change. Use of the RIN as a measure for experimental QC has recently been reported.[27,28]

Spectrophotometric analysis is often used to assess the incorporation of dyes during the labelling stage and there is often an issue with the amount of sample needed for analysis as labelling reactions are often carried out in small volumes. To check the amount of cDNA or cRNA in a reaction and assess the dye incorporation efficiency, the entire sample may be required. This means that the sample needs to be retrieved from the cuvette following analysis to be used for hybridisation. This may lead to quality issues of target contamination, if the cuvettes are not cleaned thoroughly and photobleaching of the target sample whilst measurements are being taken in the spectrophotometer. To avoid the target recycling issue, a low volume spectrophotometer such as the Nano-Drop ND-1000 UV-Vis Spectrophotometer (NanoDrop Technologies, USA) may be used.

10.2.2.4 Array Handling and Hybridisation

Due to their 'open' structure, spotted microarrays are prone to interference from scratches and dust particles on the array surface, both of which may cause unwanted fluorescence upon scanning. If a manual handling approach is adopted, differences in hybridisation efficiency or background fluorescence signal may arise across the array. This is often caused by air bubbles introduced during sample loading or by salt left over from the washing step precipitating onto the array surface during the drying procedure (see Figure 10.9). Although such problems can occur when using a hybridisation station, any person-to-person variability is removed from the experimental procedure and the hybridisation is generally more consistent across and between arrays. Analysts should be aware that differences can occur in hybridisation conditions across different hybridisation chambers, which may also have an effect on the results generated. If using an automated hybridisation and wash station, it is also important to ensure that the area of the slide where the O-ring will form the seal does not include any frosted areas or sticky labels. Such differences in surface structure or level can be enough to introduce leaks during the hybridisation and washing procedures, leading to valuable targets and arrays being lost. Many commercial array manufacturers now recommend their own hybridisation

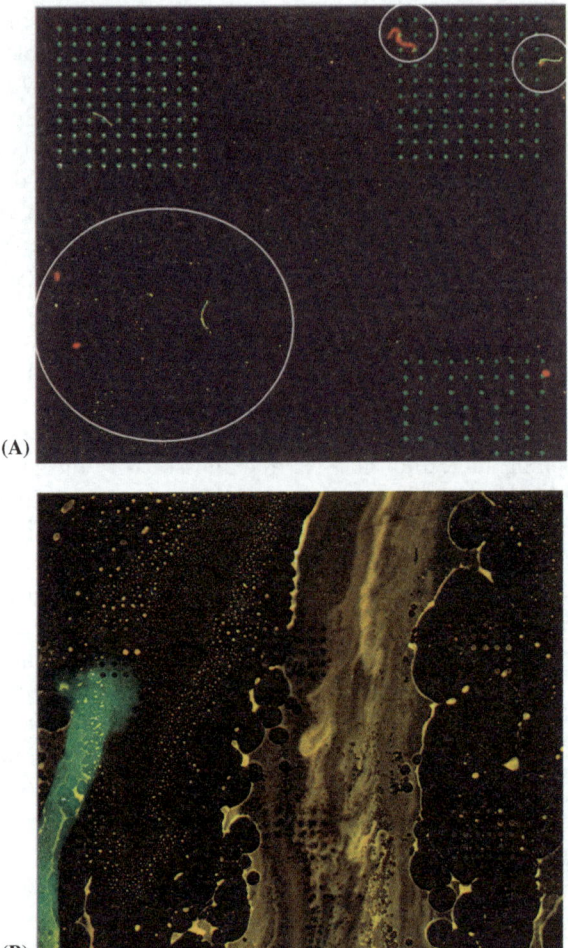

Figure 10.9 Illustration of problems that can occur during the hybridisation process. (A) The white circles highlight where dust present on the array fluoresces during a scan, interfering with the signals from the samples. This can lead to problems with array analysis. (B) Part of a slide showing how problems in hybridisation and washing can affect the fluorescent background of the slide, again leading to problems with array analysis.

protocols and even specify equipment in which hybridisation should be carried out to optimise the procedure.

10.2.2.5 Gene List Files

Most manufacturers have gene list files available specifically for commonly used image acquisition software packages available for downloading *via* their websites. This is a very good service, as it saves the operator from adapting the raw files to their own software. Nevertheless, it is important to check that these files do actually correspond to the array being used. It is always possible for mistakes to occur in these files, for example when a change of array layout is not accompanied by an update to the corresponding gene list file. Although such mistakes may be hard to spot, it is a good idea to test that the expression pattern of a subset of genes fits expectations. If in doubt, contact the manufacturer to get confirmation that the gene list file is actually the correct version.

10.2.2.6 Image Acquisition and Processing

The essential goal of image acquisition is to measure the intensity of the arrayed spots and quantify their expression levels based on these intensities. The array imaging process will also assess the reliability of the quantified spot and generate warnings for possible problems that may have occurred during the array fabrication or hybridisation stages of the process.

Spot finding is the method used to locate the signal spots in the array images and estimate the size of each spot, so that an expression level can be calculated for that probe. There are three different types of spot finding, manual, semi-automatic and automatic, based of the level of the algorithm used, and the amount of analyst intervention. In brief, for manual spot finding the computer generates a grid, which the analyst places over the spots in the image, and then each individual array spot is aligned by hand. In semi-automatic spot finding the analyst places the grid over the spots and then the computer will use algorithms for automatically adjusting the location of the grid lines. Automatic spot finding utilises advanced computer algorithms to find the spots without any human intervention.

After the spot location has been determined, a small area around the spot location can be used to quantify the spot intensity level. Then the next step in the image acquisition process is to determine which pixels in the spot location are signal and which are background. This operation is called segmentation and will not be covered in this chapter. Once this has been carried out, intensity values for each spot will be assigned.

Following image acquisition, image processing can take place. A variety of programmes are available for performing the image processing stage of an analysis, ranging from freely available open source tools to commercially produced and supported software. Using integrated packages, such as GenePix Pro (Molecular Devices, USA[29]) for image processing and Acuity (Molecular Devices, USA) for normalisation, does have certain advantages because the uniformity of data handling ensures that column headings and flags are automatically recognised from one package to the next. Although it is totally viable to use a combination of different packages, both free and commercial, for image processing and normalisation, one should be aware that it is often necessary to manipulate the format of results to fit a particular programme. Each package often produces data in a slightly different way and although they may appear to have the same column heading, for example 'ROW', they may not actually contain the same information. This can, therefore, mean that great care and advanced planning may be required to ensure consistency. Sufficient time should be allocated to perform these checks, as errors can occur when having to change columns and manipulate thousands of rows from large data sets.

Regardless of software choice, it is vitally important to the quality of data and conclusions drawn that operators are sufficiently trained in the use and interpretation of the software and data. This is another consideration to be taken into account when choosing a software package, as many commercial software manufacturers supply training and software support, including updates, whereas these tend not to be available with freeware. Sufficient time for training should also be incorporated into experimental plans in order to have confidence in data processing. It is also important to appreciate that freeware can be edited by anyone, although strict policing should keep version release controlled.

10.2.2.7 Normalisation

In order to compare microarray data, normalisation is performed to account for numerous sources of systematic variation. Such factors include labelling efficiency differences between the two fluorescent dyes, the variation in amounts of RNA labelled between the two samples, scanner settings, slide batches and dye bias depending on spot location for instance. There are a number of different normalisation methods, although no one current method is able to address all types of variation. Inappropriate normalisation can lead to incorrect conclusions or unacceptably high false positive or false negative results.

Ratio-based normalisation (global) is generally the minimum level that is required and is still widely used, although there is evidence that spatial or intensity dependent dye bias is a problem. It is based on the premise that most genes on the array are not differentially expressed and therefore the calculated mean of the ratio of medians, of every feature, should be equal to one. Other methods include normalising the data to a set of features, such as housekeeping genes, and Lowess normalisation (locally weighted scatterplot smoothing), which is a popular choice. The most suitable normalisation technique remains an open question as no one method will correct all types of variation, and is dependent on the type of array experiment being performed and the genes of interest. For example, genes expressed at or near the limits of detection would be treated differently from those that are intense and highly expressed. Even the same method can be performed in a number of ways. For example, Lowess normalisation can be done on a print-tip basis (that is, block by block) or across the whole array. It should be noted that the normalisation strategy used will affect downstream results (see Figure 10.10).

10.2.2.8 Critical Data Assessment

The normalised data set is then analysed to determine the level of expression of each of the loci tested. When using gene expression arrays, the process typically generates a list of expressed genes with the fold-change in expression relative to

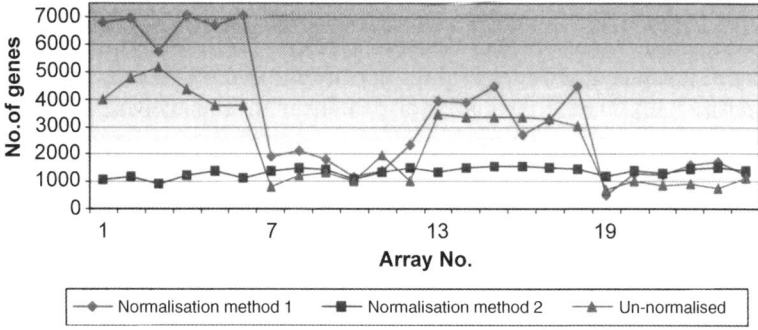

Figure 10.10 Graph showing the effect of different normalisation strategies on the number of genes considered expressed.[69] Two different strategies were used to normalise data from 24 identical arrays from one manufacturer which had undergone the same experimental protocol. Theoretically the same number of genes should be up or down regulated across all the arrays, however normalisation strategy has a large impact on the number of genes determined to be expressed.

one or more reference loci, and an associated probability value indicating the confidence in the result. Often a cut-off value is applied during the data analysis, which draws a line between background noise in the array and a level of signal which is considered to be indicative of gene expression for each spot/ locus. The probabilities values are used to filter out data with low confidence and the loci are usually ranked in order of gene expression levels, to allow clusters of genes with similar expression levels to be identified.

10.2.2.9 Drawing Biological Conclusions

The next stage in the analysis after normalising, filtering and ranking the data is to look for meaningful patterns to understand the underlying biology. Many levels of data analysis may be used to reach biologically relevant conclusions ranging from direct interpretation of the normalised data set to more complex analysis.

Data clustering is one form of downstream analysis of data to group together genes that show similar expression profiles across a number of experiments. Cluster analysis can help to establish functionally related groups of genes or predict the biochemical and physiological roles of previously uncharacterised genes. The main types of data clustering are hierarchical clustering, K-means clustering, K-medians clustering, self-organised clustering, gene sharing and principle component analysis.[30]

10.2.2.10 Data Management

Commercially available arrays contain tens of thousands of features, often in excess of 40 000, and these figures are ever increasing as printing technology improves. This means that even a single array can produce a vast number of data points, both 'raw' and processed. It is, therefore, imperative that thought is given to data management. Even small projects and low-throughput laboratories will produce quantities of data that traditional office programmes, such as Microsoft Excel, are incapable of dealing with and will need to consider storage capacity and computer processing power due to the sheer volume of data generated.

For large projects a system such as LIMS (Laboratory Information Management System) would provide an effective solution, where information is tracked from start to finish. Developing such a platform requires specialist knowledge and use of commercial packages may be expensive, although freeware such as the Bio-Array Software Environment (BASE) database system is also available.[31] Given the volume of data, creating a logical systematic naming system for files is important, particularly as there are many stages of data production such as scanning, image analysis, normalisation, and then statistical investigation. For example, a file naming system could be created that contains information such as the original scanned image, operator, scanner used, slide number, data acquisition package and normalisation strategy.

10.2.3 Technology Solutions

The scientific community is beginning to recognise the need for appropriate standards and a thorough validation of these emerging technologies, particularly for diagnostic applications. The UK Department of Trade and Industry's National Measurement System (NMS) has identified comparability of microarray measurements as being a major issue affecting full-scale exploitation of data obtained from gene expression experiments. To gain regulatory acceptance the tools used to measure gene expression must be analysed for scientific value, robustness and consistency of results across competing platforms and technologies. In particular, the FDA has highlighted a need for consistent, scientifically based, analytical guidelines to use the new tools appropriately to enable an assessment of genomic expression data, irrespective of the analytical platform. This was outlined in the FDA's critical path initiative[32] that has identified array standards as one of the most important evaluation tools that needs to be developed to enable greater uptake of and confidence in microarray-based analyses and remove barriers to further development in the microarray field. The FDA has also highlighted a need to develop an interactive, transparent process to identify true biological differences, and develop performance standards and processes to turn gene expression data into knowledge that reflects scientific consensus.

10.3 Current Commercial Microarray Quality Controls

Many of the problems described in the microarray field are the result of few standards and little method standardisation. Although this is still the case and there is currently no consensus of opinion on standardisation, manufacturers have started to fill the gap in the market, providing various controls and standardisation reagents to the microarray community. Currently these controls fall into three broad categories:

- Printing verification;
- Universal reference samples;
- Spike-in controls.

10.3.1 Printing Controls

The quality and reproducibility of printed microarray slides can directly impact on the results obtained and the conclusions drawn from microarray experiments. Variation in spot morphology, spot DNA concentration and hybridisation availability of DNA in spots all affect the downstream analysis. The production of microarray slides therefore needs very careful monitoring. In recent years, the printing technology has advanced and the quality control (QC) systems used to check commercial arrays are very stringent. Manufacturers all have their own in-house QC processes highlighting problems in printing, thus ensuring that sub-standard arrays are not released to the customer. Two examples of manufacturers' in-house QC processes are the Agilent SurePrint Technology[33] and the Codelink Gene Expression Bioarray System.[34] Although these systems work *via* different methods, one is a spotted array and the other is *in-situ* synthesis, so the QC will be different for each, they both have processes in place which ensure a very high quality of printed arrays.

For scientists printing their own microarrays, there are several products on the market to aid the quality control of the printing process. The first of these helps to look at spot morphology and highlights missing spots on an array. Arrays are randomly selected from the printing batch and the quality and presence of the spots tested using a DNA binding dye, such as SYBR® Green II or SYBR® 555.[35] The stain binds to any DNA present on the slides, thus allowing visualisation of the spots and quality assessment of the printing batch. Some kits such as the PARAGON™ DNA QC microarray stain kit[35] (Molecular Probes) also include control microarray slides useful for determining whether the staining technique itself is working, giving an added level of confidence in the results obtained from the staining.

The second type of product available for testing printing quality is fluorescently labelled oligonucleotides. These help to assess the amount of DNA available for hybridisation in each spot from randomly chosen arrays in a single printing batch. Commercial array manufacturers may use a specific oligonucleotide which binds to a single specific sequence present on all the PCR products or oligonucleotide probes printed on their arrays.[36] In-house slides can also be tested using fluorescently labelled random oligonucleotide mixes, such as Panomer™ 9 Random Oligodeoxynucleotides from Molecular Probes.[37]

At least one kit on the market (Molecular Probes) also allows the judgement of print quality, spot morphology and hybridisation efficiency using a fluorescently labelled genomic DNA mix for human expression arrays.[38]

The Full Moon system for array scanner evaluation is designed to examine, verify and calibrate a microarray scanner's performance. There are three products, an all purpose evaluation slide, a scanner calibration slide and a scanner validation slide, which are designed to quantify scanners' output and to verify general performance.[39]

10.3.2 Universal Reference RNA

Many manufacturers now provide universal reference RNAs for use in standardisation of microarray experiments.[40] These contain high-quality total RNA pooled from many cell lines of the same species, in order to give optimal broad gene coverage. They are currently commercially available for the human, rat and mouse genomes.[41–43] These controls are very useful in two-colour microarray applications, when comparing between several samples. Due to their genome-wide coverage the universal reference sample will generate a fluorescent signal across the vast majority of probes on the microarray, thus providing a base level against which the relative abundance of the transcripts from the test samples can be measured. This also allows for the comparison of data across different experiments, laboratories and technology platforms.

Universal RNA controls are also a useful tool in optimisation of both one- and two-colour microarray experiments. The commercially available reference RNA can be used as labelled template, therefore allowing other experimental parameters involved in hybridisation and detection to be investigated and optimised without wasting precious experimental RNA samples.

The use of such biologically derived RNA controls is considered to be suboptimal by some scientists due to loss of low abundance transcripts in the pooling of several cell lines, variation in cell lines over time leading to batch-to-batch variation and the inherent difficulties of working with and reliably measuring absolute qualities of the raw material.[44] The commercial companies providing such universal reference RNAs are only too aware of these problems and have vigorous quality control mechanisms in place to minimise variation between batches, as well as providing industrial scale lot sizes. It is, therefore, possible to obtain RNA for a series of experiments from the same batch, thus minimising one aspect of experimental variation.

Other RNA controls are being investigated by scientists in the field in order to address the problems with 'real' RNAs. *In vitro* transcription of cRNA provides an alternative approach for providing reference RNA samples,[45] as do reference samples derived from genomic DNA.[38,46] Some providers/manufacturers are also looking into reverse complementary oligonucleotides as a plausible alternative, as they are stable and more straightforward to quality control and quantify.[47] Such alternatives are not currently available commercially, but are possibilities for the future.

10.3.3 Spike-in Controls

Unlike the universal reference samples, which can be used to compare data across different experiments and platforms, the spike-in controls are principally for evaluation of experimental performance of expression arrays within a particular platform. They comprise exogenous RNA transcripts and are added to the experimental RNA samples prior to reverse transcription and labelling.

Several of the commercially available array validation systems, such as SpotReport™ (Stratagene),[48] ArrayControl™ (Ambion)[49] and Universal Score-Card™ (GE Healthcare)[50] contain ready-to-spot or lyophilised PCR product or oligonucleotide spotting samples and other controls that can be spotted onto an array alongside the scientist's own array probes, as well as the corresponding RNA spikes for exogenous labelled cDNA generation.[51] Other available RNA spike-in controls, such as the Two-Colour RNA Spike-in Kit from Agilent Technologies, consist of manufacturer-specific RNA transcripts designed to hybridise to the control probes present on microarrays commercially available from the same manufacturer.[52]

If dual-colour experiments are performed, different amounts of each exogenous transcript may be added to the test and reference sample to generate pre-determined signal ratios between fluorescent dyes upon scanning of the array. This information may be used to adjust for inherent differences in dye intensities during the data-normalisation procedure. Spike-in controls are also used to evaluate reverse transcription and labelling procedures, validation of spot quality and hybridisation consistency, the dynamic range of the assay, as well as comparing different hybridisation experiments.

10.4 Microarray Standardisation Initiatives

Standards in data annotation and reporting are also being developed by many standardisation initiatives ongoing throughout the microarray community. To help realise the full potential of microarrays, further development of reference materials, analytical 'best practice' guidelines and standardised approaches to experimental design and execution, performance and analysis are still required. The adoption of uniform methods and the use of standardised techniques will facilitate the production of consistently higher quality data, enable more consistent quality assurance (QA) and quality control (QC) procedures to be performed and will also facilitate the development of quality metrics to objectively assess the performance of a microarray experiment and the quality of the data generated.

Many of the current initiatives are US led, although they comprise many parties from around the world, and involve large collaborative groups of key stakeholders such as core array facilities, array manufacturers, government laboratories and regulatory agencies. Several current initiatives are described below.

10.4.1 Microarray Gene Expression Data Society (MGED)

MGED was founded in 1999 by many of the major microarray users and developers at the time, including Stanford University, Affymetrix and the European Bioinformatics Institute (EBI). Its aim is to facilitate the sharing of microarray data generated by functional genomics and proteomics experiments. MGED has many working groups all focused on different aspects of this goal, from promoting the adoption of, and further development of, MGED standards, to assisting with the development of compatible software.

The MGED society currently has two defined standards available to microarray users, with others under development. One of these is the MIAME guidelines. These guidelines outline the minimum information about the microarray experiment that should be reported in order to enable unambiguous interpretation of results and reproduction of the experiment and were first published in *Nature* in December 2001.[53] In 2002 several major scientific journals accepted the MIAME guidelines as a necessity for publication of microarray experiments. There are also several public repositories around the world designed to accept, hold and distribute MIAME compliant microarray data, including ArrayExpress at the EBI and Gene Expression Omnibus (GEO) at the National Center for Biotechnology Information (NCBI).

The second of the MGED standards is MicroArray Gene Expression (MAGE). MAGE was developed to improve the standardisation of microarray data storage and exchange; it consists of software and tools to improve sharing and transfer of microarray data between all microarray data producers and users by providing a common platform for data handling.

Uptake and adherence of these guidelines by the microarray community will undoubtedly aid the cross comparison of microarray data and help drive forward the process of standardisation in the field. An easier exchange of array data containing adequate information regarding the experimental detail will enable the biological community to derive more meaning and benefit from the many microarray studies being conducted today.

10.4.2 External RNA Control Consortium (ERCC)

The ERCC is a predominantly US-led initiative with members from over 50 private, public and academic organisations. Its aim is to produce commonly agreed upon and tested controls for use in expression assays as a true industry-wide standard control. The ERCC, founded in 2003, is a volunteer organisation that is open to anyone with an interest in working to achieve the stated goals.[54,55] These goals include:

- The production of Certified Reference Materials (CRM) qualified by the US National Institute of Standards and Technology (NIST). These will be a reference set of approximately 100 well-characterised clones of RNA transcripts from random unique sequences;

- Publication of all CRM sequences and test data demonstrating their performance across platforms;
- Protocols detailing the preparation and use of the standard controls;
- Suitable algorithms and bioinformatics tools for their quantitative assessment and evaluation.

Currently many controls used by laboratories when performing gene expression analysis experiments are custom controls, specific to assays or platforms. This makes them of limited use as industry-wide controls to aid comparison across laboratories and platforms. The standards being developed by the ERCC are intended to be useful for evaluating the technical performance at many stages of gene expression experiments, such as QC of sample collection and sample labelling, platform characterisation, comparison and optimisation.

10.4.3 Microarray Quality Control (MAQC) Project

The MAQC project, started in 2005, is a US-led initiative involving FDA centres, major providers of microarray platforms and RNA samples, US Environmental Protection Agency (EPA), NIST and many other stakeholders, both academic and industrial. The aim of the project is to establish thresholds and quality control metrics, which will allow the objective assessment of achievable performance for many microarray platforms and the merits of various data analysis methods on data sets generated from differing microarray platforms, as well as readily accessible reference RNA samples.

Two RNA samples from each of three species have been analysed in gene expression experiments using microarrays and other technologies (such as qRT-PCR) and the precision and cross platform/laboratory comparability of the data sets is being assessed. This project has generated a large controlled data set which is publicly available for other laboratories to analyse alongside their own data. As well as the data sets and RNA standards, the project has produced guidance on microarray quality control and data analysis.[18] All of the results of the MAQC project are published and the data are now available.[56-65]

This availability of calibrated RNA samples combined with the reference data sets produced and the quality control and data analysis guidelines allows individual laboratories to identify and correct procedural failures more easily. This in turn should help to improve the results obtained using microarray technology and the successful and reliable application of the technology in many fields, such as pharmacogenomics and toxicogenomics.

10.4.4 Association of Biomolecular Research Facilities (ABRF) Microarray Research Group (MARG)

The ABRF[66] is another US initiative, the main focus of which is to promote communication and cooperation among core laboratories providing microarray and data analysis services. It also conducts studies aiming to assess technological advancements in the field of microarrays. MARG disseminates the findings from these studies to interested parties through publication in the literature, enabling end users to make best use of the available technology to achieve high quality results. The MARG is also working to provide ways of sharing information relevant to the administration of facility laboratories that provide microarray technologies as a shared resource.[67]

10.4.5 Measurements for Biotechnology (MfB) Programme

The MfB programme is funded by the UK Government's Department of Trade and Industry (DTI) and is just one of the programmes supporting the development of the UK's National Measurement System. The National Measurement System is the technical and organisational infrastructure which ensures a consistent and internationally recognised basis for measurement within the UK. The MfB programme aims to improve the accuracy and reliability of biomeasurements important for industry, strengthen the measurement science underpinning the regulatory regime for biotechnology and ensure that the UK biomeasurement system is co-ordinated and developed in harmony with those of other countries. A series of initiatives have previously been and are currently being led by LGC Limited (UK) to address genomic standardisation issues, under the MfB gene measurement theme. Projects such as this can only help to improve the comparability of data sets in the microarray arena and eventually improve the standardisation of microarray experiments. Further information on all these initiatives can be found on the MfB website.[68]

10.4.5.1 Specificity Standards and Performance Indicators

One current initiative is concerned with the development of 'best practice' protocols, reference standards and toolkits to increase confidence in array technologies.

A challenging part of any microarray experiment is to limit potential systematic variation that may be introduced. Implementing suitable standards and performance indicators should make it possible to monitor assay performance. Generic control sequences that indicate the discriminatory power of the system, when incorporated into array-based assays, will increase confidence in the performance and comparability of these technologies. In doing so, one may be able to attribute those factors having a more weighted effect on the performance of an array, which in turn may assist in the development, optimisation and validation of array based assays and facilitate the ability to compare data between experiments, platforms and laboratories. An added challenge for array-based Single Nucleotide Polymorphism (SNP) genotyping assays, which rely on the differential hybridisation properties of matched and mismatched DNA sequences, is to use assay conditions that are discriminatory for all sequences present. Generic control sequences incorporated into assays that indicate the discriminatory power of the system will increase confidence in the performance, quality of data and comparability of the varied technologies.

Microarrays 233

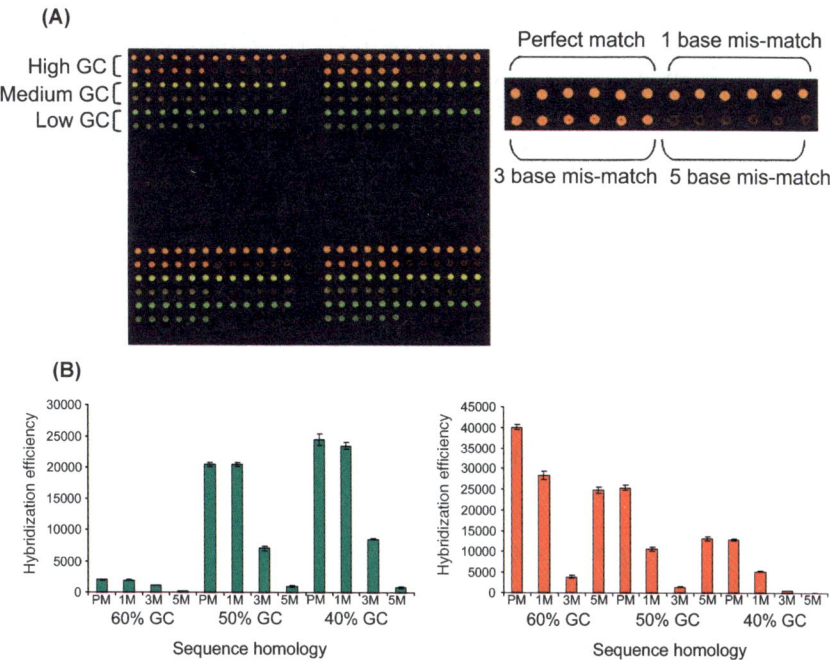

Figure 10.11 Typical results utilising the LGC array performance indicators. (A) Image of an array of the performance control indicators hybridised with perfectly matched reverse complement. Each block comprises a 12 × 8 feature format with each oligo spotted in replicates of six within each block. The enlarged slide area shows the six replicates of each of the perfectly matched (PM), single base mis-matched (1M), 3 base mismatched (3M) and 5 base mis-matched (5M) probes for the high GC (60% GC) sequence after hybridisation. The high GC, medium GC (50%) and low GC (40%) sequence reverse complements were labelled with both Cy3 (green) and Cy5 (red), and added to the hybridisation at varying concentrations. Specifically, the Cy3 labelled complements were added at a constant concentration of 0.1 pM, while the Cy 5 labelled target concentrations varied between 0.05 and 0.2 pM, giving Cy5:Cy3 ratios of: 0.5:1, 1:1 and 2:1 for the 40%, 50% and 60% probes respectively. This is reflected in the differing colouration of the 40–60% probes on the slide (green – red). (B) Graphs showing the median fluorescence intensity (MFI) of Cy3 and Cy5 signals for all replicate spots on the array. As expected the perfectly matched probes generally exhibit higher fluorescence and hybridisation efficiency, and the mismatches have a greater destabilising effect on the lower GC probes. The characteristic MFI profiles may be used to assess the hybridisation efficiency and discrimination of an assay.

A set of reference standards has therefore been designed by LGC as part of the MfB programme (2000–2003) (Figure 10.11A). The standards are three sets of 'spike-in' fluorescently labelled oligonucleotide probes of high, medium and low melting temperature (T_m), designed to hybridise to panels of probes incorporated in the array. Each set contains four probes with varying numbers of base mismatches from a core reference sequence. Labelled reverse complementary oligonucleotide, designed to the reference sequence in each set, is included in the hybridisation mix. Comparison of the same standards printed in different places within the same slide or across different slides can be used as indicators of intra- or inter-slide hybridisation efficiency respectively. Determination of the relative signal intensities for the standards within each set provides an overall indication of hybridisation specificity for the reaction conditions used (Figure 10.11B). The probe sets are also used for array quality control by monitoring the deviation of the standards from an average profile across the array. This allows for both intra- and inter-array comparisons and thus the reproducibility across experiments can also be monitored. All this will aid in the optimisation, validation and comparability of array-based assays. Spatial, pin-specific and spot deposition effects can also be identified and detected using the standards spotted in replicate across the array. These materials[68] complement the ERCC standards discussed in Section 10.4.2.

10.4.5.2 Comparability of Gene Measurements

Creating reproducible data with a high level of consistency across experiments and various platforms is still widely accepted by the scientific community as a major problem, as discussed in earlier sections. A previously completed MfB project[69] aimed to address some of these concerns by producing a large, controlled data set allowing the statistical assessment of the comparability of gene expression measurements performed on different microarray systems, as well as enabling an evaluation of the effects of different parts of the microarray process. The major sources of variability could, therefore, be identified, showing which may have the most effect on the data output and potentially on the biological interpretation of the data.[68]

10.4.5.3 Quality Metrics/Increasing Confidence in Toxicogenomic Measurement

LGC is working with the European Bioinformatics Institute (EBI) under the MfB programme (2004–2007) to gain a greater understanding of the impact of experimental variables on the conclusions drawn from toxicogenomic data, and to develop a framework for microarray standardisation in order to maximise the potential of the technology. In the area of toxicology the use of genomic approaches (toxicogenomics) using technologies such as microarrays promises a substantial impact across the entire drug discovery and development pipeline. Building on initiatives instigated in the previous MfB programme (2000–2003) to improve confidence in microarray based measurements, work is underway to develop a panel of 'quality metrics'. These are a set of factors which will help in assessing the quality of data from a microarray experiment, therefore providing objective performance measurements for validating and standardising toxicogenomic arrays.

The large data set produced during the previous MfB programme (2000–2003) (see Section 10.4.5.2) has been uploaded to the EBI's public data repository, Array Express, as part of this initiative. Array data from a model toxicology system using a chemical with a known mode of action is being used to develop and validate the quality metrics and this second data set is also publicly available *via* ArrayExpress. It is anticipated that this second data set may also be developed into a training tool to allow users to validate data analysis approaches for identifying consistent and reproducible gene expression changes. It is hoped that making these valuable data sets easily accessible will aid microarray validation, and help determine the reproducibility of gene expression profiles between replicate arrays and across arrays produced by different manufacturers.

As well as developing a set of quality metrics, an assessment of the impact of starting RNA quality and the consistency of toxicogenomic responses measured on different array platforms will be key outputs of the initiative.

10.4.5.4 Standard Units to Measure Gene Expression

The MfB Gene Expression Units Working Group was established as part of an initiative to develop a standard approach for measuring gene expression, to include reference methods, materials and units. The objective of the working group is to recommend approaches for better standardisation of procedures used in gene expression measurements through discussions and consultations. Practical assessments of recommended approaches and the development of associated methodologies, standards and reference materials are among the aims of this work.[68,70]

10.5 Summary

Microarrays have the ability to revolutionise analytical applications in many fields including clinical and medical. If, however, the full potential of this technology is to be realised and move from being a research utensil to a fully validated, robust, analytical tool, it is recognised by the scientific community that more concerted efforts to develop universally applicable standards, reference materials and analytical guidelines to assist in the comparability and standardisation of microarray experiments, such as those described in this chapter, are required. There is, in fact, a need to standardise the standards to ensure universal and consistent uptake of the procedures.[71]

References

1. M. Schena, D. Shalon, R. W. Davis and P. O. Brown, *Science*, 1995, **270**, 467.
2. U. Maskos and E. M. Southern, *Nucleic Acids Res.*, 1992, **20**, 1679.
3. S. Venkatasubbarao, *Trends Biotechnol.*, 2004, **22**, 630.

4. S. A. Dunbar, *Clin. Chim. Acta*, 2006, **363**, 71.
5. D. D. Shoemaker, D. A. Lashkari, D. Morris, M. Mittmann and R. W. Davis, *Nat. Genet.*, 1996, **14**, 450.
6. R. J. Fulton, R. L. McDade, P. L. Smith, L. J. Kienker and J. R. Kettman, Jr., *Clin. Chem. (Washington, DC, USA)*, 1997, **43**, 1749.
7. P. Angenendt, *Drug Discov. Today*, 2005, **10**, 503.
8. M. Uttamchandani, J. Wang and S. Q. Yao, *Mol. Biosyst.*, 2006, **2**, 58.
9. V. Mello-Coelho and K. L. Hess, *Braz. J. Med. Biol. Res.*, 2005, **38**, 1543.
10. T. J. Aitman, *Brit. Med. J.*, 2001, **323**, 611.
11. R. B. Stoughton, *Annu. Rev. Biochem.*, 2005, **74**, 53.
12. T. Watanabe, Y. Komuro, T. Kiyomatsu, T. Kanazawa, Y. Kazama, J. Tanaka, T. Tanaka, Y. Yamamoto, M. Shirane, T. Muto and H. Nagawa, *Cancer Res.*, 2006, **66**, 3370.
13. S. Wang and Q. Cheng, *Methods Mol. Biol.*, 2006, **316**, 49.
14. F. De Smet, N. L. Pochet, K. Engelen, T. Van Gorp, P. Van Hummelen, K. Marchal, F. Amant, D. Timmerman, B. L. De Moor and I. B. Vergote, *Int. J. Gynecol. Cancer*, 2006, **16**, Suppl 1, 147.
15. http://www.roche-diagnostics.com/press_lounge/press_releases/archive/pr_42303.html (accessed July 2007).
16. J. de Leon, *Expert Rev. Mol. Diagn.*, 2006, **6**, 277.
17. A. M. Glas, A. Floore, L. J. Delahaye, A. T. Witteveen, R. C. Pover, N. Bakx, J. S. Lahti-Domenici, T. J. Bruinsma, M. O. Warmoes, R. Bernards, L. F. Wessels and L. J. Van't Veer, *B.M.C. Genomics*, 2006, **7**, 278.
18. L. Shi, W. Tong, Z. Su, T. Han, J. Han, R. K. Puri, H. Fang, F. W. Frueh, F. M. Goodsaid, L. Guo, W. S. Branham, J. J. Chen, Z. A. Xu, S. C. Harris, H. Hong, Q. Xie, R. G. Perkins and J. C. Fuscoe, *BMC Bioinformatics*, 2005, **6**(Suppl 2), S11.
19. C. L. Archard, M. Cutriss and C. A. Foy, 2003, *LGC Report*, *2003*, **LGC/VAM/2003/051**, available at http://www.nmschembio.org.uk/ (accessed July 2007).
20. M. T. Andersen, M. Burns, D. Hopkins and C. A. Foy, *LGC Report*, 2004, **LGC/MFB/2004/015**, available at www.nmschembio.org.uk with free registration.
21. M. Reimers, *Addict. Biol.*, 2005, **10**, 23.
22. M. L. Lee, F. C. Kuo, G. A. Whitmore and J. Sklar, *Proc. Natl. Acad. Sci. USA*, 2000, **97**, 9834.
23. A. B. Goryachev, P. F. Macgregor and A. M. Edwards, *J. Comput. Biol.*, 2001, **8**, 443.
24. R. Simon, M. D. Radmacher and K. Dobbin, *Genet. Epidemiol.*, 2002, **23**, 21.
25. K. Dobbin, J. H. Shih and R. Simon, *J. Natl. Canc. Inst.*, 2003, **95**, 1362.
26. O. Mueller, S. Lightfoot and A. Schroeder, 2004, at http://www.chem.agilent.com/temp/radA4957/00001064.PDF (accessed October 2007).
27. L. Jones, G. Hughes, A. D. Strand, F. Collin, S. B. Dunnett, C. Kooperberg, A. Aragaki, J. M. Olson, S. J. Augood, R. L. Faull, R. Luthi-Carter, V. Moskvina and A. K. Hodges, *BMC Bioinformatics*, 2006, **7**, 211.

28. A. Schroeder, O. Mueller, S. Stocker, R. Salowsky, M. Leiber, M. Gassmann, S. Lightfoot, W. Menzel, M. Granzow and T. Ragg, *BMC Mol. Biol.*, 2006, **7**, 3.
29. http://www.moleculardevices.com/ (accessed July 2007).
30. Y. X. Zhou, P. Kalocsai, J. Chen, S. Shams, in *Microarray Biochip Technology*, ed. M. Schena, Eaton Publishing, Westborough, MA, 2000, p. 167, ISBN 978-1-881299-37-0.
31. http://base.thep.lu.se/#License (accessed July 2007).
32. US Food and Drug Administration Critical Path Initiative, at http://www.fda.gov/oc/initiatives/criticalpath/ (accessed July 2007).
33. Agilent SurePrint technology at, http://www.chem.agilent.com/Scripts/Generic.ASP?lPage=557&indcol=N&prodcol=Y (accessed July 2007).
34. Codelink Gene Expression Bioarray System at, http://www6.amershambiosciences.com/aptrix/upp01077.nsf/Content/codelink_gene_expression#Quality (accessed July 2007).
35. PARAGON DNA microarray kit at, http://probes.invitrogen.com/media/pis/mp32930.pdf (accessed July 2007).
36. Clontech BD Atlas™ Custom Array Printing Services, *Clontechniques*, 2002, **XVII**, 4, at http://www.clontech.com/upload/images/ctq/full/CTQOCT02.pdf (accessed July 2007).
37. Panomer product information at, http://probes.invitrogen.com/media/pis/mp21678.pdf (accessed July 2007).
38. PARAGON DNA hybridisation kit at, http://probes.invitrogen.com/media/pis/mp32938.pdf (accessed July 2007).
39. http://www.fullmoonbio.com/ (accessed July 2007).
40. N. Novoradovskaya, M. L. Whitfield, L. S. Basehore, A. Novoradovsky, R. Pesich, J. Usary, M. Karaca, W. K. Wong, O. Aprelikova, M. Fero, C. M. Perou, D. Botstein and J. Braman, *BMC Genomics*, 2004, **5**, 20.
41. Strategene Universal Human Reference RNA at, http://ngfn.rzpd.de/uploads/BMuignbTOiEtM3VdTXNxbg/w-dEurGM0PURtfL8R2Iyxg/RNA_common_reference_.pdf (accessed July 2007).
42. BD Clontech™ Universal Reference total RNA, *Clontechniques*, 2003, **XVIII**, 18, at http://www.clontech.com/upload/images/ctq/full/CTQAPR03.pdf (accessed July 2007).
43. Stratagene Universal Mouse Reference RNA, at http://www.stratagene.com/manuals/740100.pdf (accessed May 2007).
44. M. T. Andersen and C. A. Foy, *Anal. Bioanal. Chem.*, 2004, **381**, 87.
45. E. Sterrenburg, R. Turk, J. M. Boer, G. B. van Ommen and J. T. den Dunnen, *Nucleic Acids Res.*, 2002, **30**, 1.
46. Y. Wei, J. M. Lee, C. Richmond, F. R. Blattner, J. A. Rafalski and R. A. LaRossa, *J. Bacteriol.*, 2001, **183**, 545.
47. A. M. Dudley, J. Aach, M. A. Steffen and G. M. Church, *Proc. Natl. Acad. Sci. USA*, 2002, **99**, 7554.
48. Stratagene. SpotReport(R)-3 Array Validation System Instruction Manual, at http://www.stratagene.com/manuals/252005.pdf.

49. Ambion. ArrayControl(TM) Instruction Manual, at http://www.ambion.com/techlib/prot/fm_1780.pdf (accessed July 2007).
50. GE Healthcare Universal Scorecard, at http://www5.gelifesciences.com/aptrix/upp01077.nsf/Content/Products?OpenDocument&parentid=460766&moduleid=165076&zone=.
51. Y. Wu, P. de Kievit, L. Vahlkamp, D. Pijnenburg, M. Smit, M. Dankers, D. Melchers, M. Stax, P. J. Boender, C. Ingham, N. Bastiaensen, R. de Wijn, D. van Alewijk, H. van Damme, A. K. Raap, A. B. Chan and R. van Beuningen, *Nucleic Acids Res.*, 2004, **32**, e123.
52. Agilent Two Color RNA Spike-in Kit Protocol, 2005, Publication number 5188-5279, at http://www.chem.agilent.com/scripts/literaturePDF.asp?iWHID=40485 (accessed July 2007).
53. A. Brazma, P. Hingamp, J. Quackenbush, G. Sherlock, P. Spellman, C. Stoeckert, J. Aach, W. Ansorge, C. A. Ball, H. C. Causton, T. Gaasterland, P. Glenisson, F. C. Holstege, I. F. Kim, V. Markowitz, J. C. Matese, H. Parkinson, A. Robinson, U. Sarkans, S. Schulze-Kremer, J. Stewart, R. Taylor, J. Vilo and M. Vingron, *Nature Genetics*, 2001, **29**, 365.
54. M. Cronin, K. Ghosh, F. Sistare, J. Quackenbush, V. Vilker and C. O'Connell, *Clin. Chem. (Washington, DC, USA)*, 2004, **50**, 1464.
55. S. C. Baker, S. R. Bauer, R. P. Beyer, J. D. Brenton, B. Bromley, J. Burrill, H. Causton, M. P. Conley, R. Elespuru, M. Fero, C. Foy, J. Fuscoe, X. Gao, D. L. Gerhold, P. Gilles, F. Goodsaid, X. Guo, J. Hackett, R. D. Hockett, P. Ikonomi, R. A. Irizarry, E. S. Kawasaki, T. Kaysser-Kranich, K. Kerr, G. Kiser, W. H. Koch, K. Y. Lee, C. Liu, Z. L. Liu, A. Lucas, C. F. Manohar, G. Miyada, Z. Modrusan, H. Parkes, R. K. Puri, L. Reid, T. B. Ryder, M. Salit, R. R. Samaha, U. Scherf, T. J. Sendera, R. A. Setterquist, L. Shi, R. Shippy, J. V. Soriano, E. A. Wagar, J. A. Warrington, M. Williams, F. Wilmer, M. Wilson, P. K. Wolber, X. Wu and R. Zadro, External RNA Controls Consortium, *Nature Methods*, 2005, **2**, 731.
56. L. Guo, E. K. Lobenhofer, C. Wang, R. Shippy, S. C. Harris, L. Zhang, N. Mei, T. Chen, D. Herman, F. M. Goodsaid, P. Hurban, K. L. Phillips, J. Xu, X. Deng, Y. A. Sun, W. Tong, Y. P. Dragan and L. Shi, *Nat. Biotechnol.*, 2006, **24**, 1162.
57. R. Shippy, S. Fulmer-Smentek, R. V. Jensen, W. D. Jones, P. K. Wolber, C. D. Johnson, P. S. Pine, C. Boysen, X. Guo, E. Chudin, Y. A. Sun, J. C. Willey, J. Thierry-Mieg, D. Thierry-Mieg, R. A. Setterquist, M. Wilson, A. B. Lucas, N. Novoradovskaya, A. Papallo, Y. Turpaz, S. C. Baker, J. A. Warrington, L. Shi and D. Herman, *Nat. Biotechnol.*, 2006, **24**, 1123.
58. W. Tong, A. B. Lucas, R. Shippy, X. Fan, H. Fang, H. Hong, M. S. Orr, T. M. Chu, X. Guo, P. J. Collins, Y. A. Sun, S. J. Wang, W. Bao, R. D. Wolfinger, S. Shchegrova, L. Guo, J. A. Warrington and L. Shi, *Nat. Biotechnol.*, 2006, **24**, 1132.
59. L. Shi, L. H. Reid, W. D. Jones, R. Shippy, J. A. Warrington, S. C. Baker, P. J. Collins, F. de Longueville, E. S. Kawasaki, K. Y. Lee, Y. Luo, Y. A. Sun, J. M. Willey, R. A. Setterquist, G. M. Fischer, W. Tong,

Y. P. Dragan, D. J. Dix, F. W. Frueh, F. M. Goodsaid, D. Herman, R. V. Jensen, C. D. Johnson, C. D. E. K. Lobenhofer, R. K. Puri, U. Schrf, J. Thierry-Mieg, C. Wang, M. Wilson, P. K. Wolber, L. Zhang and W. Slikker Jr., *Nat. Biotechnol.*, 2006, **24**, 1151.
60. R. D. Canales, Y. Luo, J. C. Willey, B. Austermiller, C. C. Barbacioru, C. Boysen, K. Hunkapiller, R. V. Jensen, C. R. Knight, K. Y. Lee, Y. Ma, B. Maqsodi, A. Papallo, E. H. Peters, K. Poulter, P. L. Ruppel, R. R. Samaha, L. Shi, W. Yang, L. Zhang and F. M. Goodsaid, *Nat. Biotechnol.*, 2006, **24**, 1115.
61. T. A. Patterson, E. K. Lobenhofer, S. B. Fulmer-Smentek, P. J. Collins, T. M. Chu, W. Bao, H. Fang, E. S. Kawasaki, J. Hager, I. R. Tikhonova, S. J. Walker, L. Zhang, P. Hurban, F. de Longueville, J. C. Fuscoe, W. Tong, L. Shi and R. D. Wolfinger, *Nat. Biotechnol.*, 2006, **24**, 1140.
62. H. Ji and R. W. Davis, *Nat. Biotechnol.*, 2006, **24**, 1112.
63. D. A. Casciano and J. Woodcock, *Nat. Biotechnol.*, 2006, **24**, 1103.
64. F. W. Frueh, *Nat. Biotechnol.*, 2006, **24**, 1105.
65. D. J. Dix, K. Gallagher, W. H. Benson, B. L. Groskinsky, J. T. McClintock, K. L. Dearfield and W. H. Farland, *Nat. Biotechnol.*, 2006, **24**, 1108.
66. Association of Biomolecular Resource Facilities, at http://www.abrf.org/ (accessed July 2007).
67. K. L. Knudtson, H. Auer, A. I. Brooks, C. Griffin, G. Grills, S. Hester, G. Khitrov, K. S. Lilley, A. Massimi, J. P. Tiesman and A. K. Viale, *J. Biomol. Tech.*, 2006, **17**, 176.
68. MfB website, http://www.mfbprog.org.uk/ (accessed July 2007).
69. D. Hopkins, M. T. Andersen, M. Burns and C. A. Foy, *LGC Report*, 2004, **LGC/MfB/2004/009**, available at www.nmschembio.org.uk through free registration.
70. M. T. Andersen and C. A. Foy, *LGC Report*, 2005, **LGC/MFB/2005/023**, available at www.nmschembio.org.uk through free registration.
71. J. Quackenbush, *Mol. Syst. Biol.*, 2006, **2**, 2006.

Subject Index

Accreditation, 13
Accuracy, 48
Acrylamide gel electrophoresis, 125–6
Agarose gel electrophoresis, 125
Amplified fragment length polymorphism, DNA requirements, 64
Analysts *see* Staff
Analyte concentration, 6–7
Analytical consistency, 6
Analytical process
  requirements, 2–3
  stages in, 3–4
Analytical requirement, 53–4
Anion exchange resins, 73
Archival tissue samples, RNA extraction, 191–2
Archiving, 36–7
  electronic data, 37
Ascorbic acid, 72
Association of Biomolecular Research Facilities Microarray Research Group, 231–2
Audit, 12
Automated analysis, 35–6

Beer-Lambert Law, 84
Best scientific practice, 13
Bias, 46–7, 56
Biological contamination, 7
Biotechnology and Biological Sciences Research Council, 13
British Standards Institute, 14

Calibration of equipment, 25
  fluorescence spectroscopy, 88–9
  ultraviolet spectroscopy, 85
Calibration standards, 15–26
Capillary electrophoresis, 126
Cation exchange resins, 73
Cell lysis, 61, 68–71
  detergent/denaturant, 68, 69
  DNA damage, 70–1
  enzymes used in, 70
  extraction buffer
    EDTA concentration, 70–1
    pH, 70
    salt concentration, 70
  inaccessibility of cells to lysis forces, 68
  lytic enzyme concentration/activity, 68–9
Certification, 13, 14
Certified reference materials, 6, 31, 52
  lack of, 8
Cetyltrimethylammonium bromide, 69
Chain of custody, 33–4
Chitinase, 70
Choice of method, 5
  DNA extraction, 62–6
Coefficient of variation, 45
Comparative genomic hybridisation, 213
Constituent phosphorus determination, 97–8
Constraints on analysis, 4
Contamination
  samples, 67
  ultraviolet spectroscopy, 87–8

Subject Index

Contamination control, 33
  PCR, 122–8
    decontamination methods, 124, 128
    dedicated equipment, 123–4
    physical laboratory separation, 123–4
    pipettes, 124
  quantitative real-time PCR, 152–3
Controls, 29
  in-house materials, 31–2
  performance, 32–3
  quantitative real-time PCR, 153
  see also Quality Control
Credibility of laboratories, 11
Critical parameters, 54
CRMs *see* Certified reference materials
Cuvettes, 85–6
Cycle sequencing, DNA requirements, 64
Cysteine, 72

Data
  criticality of, 42
  electronic
    archiving, 37
    recording, 35–6
  interpretation, 4
  quality, 8–9
  reporting, 4, 36
Degenerate oligonucleotide primed PCR, 179–80
Denaturants, 68, 69
Deoxynucleotide triphosphates, 107
Department of Environment, Food and Rural Affairs (DEFRA), 13
Detection limits, 44, 50–1
Detergents, 68, 69
Diethyl-dithiocarbamic acid, 72
Dithiothreitol, 72
DNA
  high molecular weight, 60–1
  reference standard, 90
DNA degradation, 7, 60–1
DNA extraction, 59–82
  automation of, 76–7
  cell and membrane lysis, 61, 68–71
  choice of procedure, 62–6
    history of sample, 62
    impact of methodology, 63–6
    sample composition, 62–3
    standardised techniques, 63
    subsequent analytical procedures, 63
    time and resources available, 63
  concentration/amount, 60
  concentration of DNA, 62, 74–6
  DNA purification, 60, 61–2, 71–4
  integrity, 60–1
  protection/stabilisation of released DNA, 61
  protocols for, 77–8
  sample preparation, 61, 66–7
  sample storage, 66
  separation of nucleic acids from cellular debris, 61, 67, 71
  validation issues, 66–76
DNA polymerase, 147
  thermostable, 109–10
DNA purification, 71–4
  column cleaning, 72
  extraction buffer, 72
  RNase treatment, 72–4
DNA quantification, 83–100
  applications, 83
  comparability of methods, 98–9
  constituent phosphorus determination, 97–8
  enzymatic, 93–4
  fluorescence spectroscopy, 88–93
    calibration graph, 88–9
    DNA standard, 90
    dye, 91–3
    dye concentration, 90
    dye selection, 90
    measurement conditions, 90–1
    microtitre plates, 90
    reference blank, 90
    sample preparation, 89
  polymerase chain reaction, 93
  primary methods, 94–6
    gravimetric analysis, 95
    isotope dilution mass spectrometry, 95–6

ultraviolet spectroscopy, 84–8
  calibration of spectrophotometer, 85
  cuvettes, 85–6
  extinction coefficient, 84–5
  light source, 86
  presence of contaminants, 87–8
  reference blank, 86
  sample dilution, 86
  sample preparation, 86
DNA sequencing, 212
DNA template, 105–6
  concentration, 105–6
  integrity, 105
Documentation, 38, 57
Dot/slot blot hybridisation, DNA requirements, 64

Electronic data
  archiving, 37
  recording, 35–6
Electrophoresis
  acrylamide gel, 125–6
  agarose gel, 125
  capillary, 126
  pulsed field gel, 64
Enzymatic DNA quantification, 93–4
Equipment, 24–5
  analytical requirement, 24
  calibration, 25
  instrumentation, 141–4
  log books and maintenance, 24–5
  maintenance, 24–5
  'ownership', 24
Equipment qualification, 41
Ethidium bromide, 91–2
European Commission
  Directive 99/11/EEC, 16
  Directive 99/12/EEC, 16
Experimental design, 29–30
Experimental requirements, 5
External quality assessment, 5, 28
External RNA Control Consortium, 230–1
Extinction coefficient, 84–5

Fluorescence spectroscopy, 88–93
  calibration graph, 88–9
  DNA standard, 90
  dye, 91–3
    concentration, 90
    ethidium bromide, 91–2
    Hoechst 33258, 92–3
    PicoGreen, 92
    selection, 90
    SYBR, 92
  measurement conditions, 90–1
  microtitre plates, 90
  reference blank, 90
  sample preparation, 89
Fluorescent dyes, 91–3
  ethidium bromide, 91–2
  Hoechst 33258, 92–3
  PicoGreen, 92
  SYBR range, 92
Fluorophores, 137
  choice of, 140–1
Food Standards Agency, 13
Förster resonance energy transfer, 134
Freezing of samples, 189–90

Gel filtration, 73
Gene expression profiling, 212
Gene specific PCR, DNA requirements, 64
Genetically modified foods, 8, 52
Genetically modified organisms, 83
Genotyping, 212
GLP Monitoring Authority, 16
Good Laboratory Practice, 2, 20–1, 41
Gravimetric analysis, 95
Guanidine isothiocyanate, 69
Guanidinium isothiocyanate, 190

Hoechst 33258, 92–3
Hook effect, 152
Hot-start mechanisms, PCR, 110, 112
Hybridisation probes, 136–8
Hydroxyapatite, 73
Hypochromic shift, 84

Subject Index

Independent Quality Assessment, 28–9
Inhibitors, 113–17, 194–8
In-house control materials, 31–2
Instrumentation, 141–4
  see also Equipment
Intermediate precision, 46
International Electrotechnical
    Commission, 13
Internationally recognised assessed
    standards, 12–17
International Organization for
    Standardization *see* ISO
ISO 9001:2000, 13, 14, 18
ISO 15189:2003, 13, 16
ISO/IEC 9001, 41
ISO/IEC 17025:2005, 13, 15–16, 18, 40
Isotope dilution mass spectrometry, 95–6

Joint Code of Practice for Research,
    13, 16–17

Laboratory environment, 22–4
  safety, 22–3
  spatial separation, 23–4
Laboratory Information Management
    System (LIMS), 34
Laboratory performance assessment, 5
Ligation-mediated PCR, 181
Limits of detection *see* Detection limits
Linearity, 51–2
Linear range, 56
Locally Controlled Documentation, 22
Locked nucleic acid, 171–4
Log books, 24–5
Lysozyme, 70
Lyticase, 70

Magnesium chloride, 107, 146
Maintenance of equipment, 24–5
Management systems, 10–12
  internationally recognised assessed
    standards, 12–17
  selection and implementation, 17–37
    archiving, 35–7
    equipment, 24–5

laboratory environment, 22–4
Locally Controlled Documentation,
    22
methods, 28–30
Quality Control, 3, 6, 10–39
Quality Manual, 19–20, 21
Quality Procedures, 20
reagents, 25–6
recording and reporting, 35–6
samples, 33–5
staff, 26–7
Standard Operating Procedures, 20–2
Master mix, 146–7
Matrix
  complex, 7
  degradation of, 7–8
Matrix-assisted laser desorption-
    ionisation, 127
Measurement quality
  challenges to, 6–8
    biological contamination, 7
    complex matrices, 7
    DNA degradation, 7
    lack of CRMs, 8
    limited sample availability, 8
    low analyte concentration, 6–7
    matrix degradation, 7–8
  external assessment, 5
Measurements for Biotechnology
    Programme, 232–5
  comparability of gene
    measurements, 234
  quality metrics, 234–5
  standards and performance
    indicators, 232–4
  standard units for gene expression
    measurement, 235
Measurement uncertainty, 30, 52–3
Medicines and Healthcare Products
    Regulatory Agency, 16
Melting curve analysis, 139–40
β-Mercaptoethanol, 72
Methods, 28–30
  documentation, 28
  expected level of utlisation, 43

experimental design, 29–30
fitness for purpose, 28
Independent Quality Assessment, 28–9
measurement uncertainty, 30
metrological traceability, 28
robustness of, 42–3
validation, 2, 4, 29
Method performance parameters, 43–53
  accuracy, 48
  bias, 46–7
  detection limit (sensitivity), 50–1
  measurement uncertainty, 52–3
  precision, 44–6
  recovery, 47–8
  ruggedness (robustness) testing, 49
  selectivity, 49–50
  working range and linearity, 51–2
Metrological traceability, 10
Microarray Gene Expression Data Society, 230
Microarrays, 208–39
  applications, 212–14
  biological conclusions, 225
  critical data assessment, 224–5
  current problems, 214–16
  data management, 225–6
  definition of, 208–10
  experimental design, 217–19
  gene list files, 222–3
  handling and hybridisation, 221–2
  image acquisition and processing, 223–4
  layout and content, 219
  nomenclature, 210
  normlisation, 224
  quality control, 226–9
    printing, 227–8
    spike-in controls, 229
    universal reference RNA, 228
  standardisation initiatives, 229–35
    Association of Biomolecular Research Facilities Microarray Research Group, 231–2
    External RNA Control Consortium, 230–1
    Measurements for Biotechnology Programme, 232–5
    Microarray Gene Expression Data Society, 230
  target quality, 220–1
  technology solutions, 226
  types of, 210–12
Microtitre plates, 90
Mis-priming, 176
Modde software, 43
Molecular beacons, 135–6
Multiple displacement amplification, 180–1
Multiplex PCR, 167–78
  advantages and disadvantages, 177–8
  amplification target, 169
  applications, 177, 178
  design, 168–9
  detection strategies, 177
  DNA requirement, 64
  mis-priming events, 176
  number of targets amplified, 168
  optimisation, 174–6
    cycling parameters, 175–6
    initial assay development, 174
    reaction components, 174–5
  primer design, 170–1
  primer positioning, 169–70
  specificity, 176–7
  standardization of oligonucleotide Tm, 171–4
    base analogues, 171
    locked nucleic acid, 171–4
    peptide nucleic acid, 171
  untemplated nucleotide addition, 177
MuPlex system, 168
MVAL, 43

Natural Environmental Research Council, 13
5' Nuclease assay, 134–5

Organisation for Economic Co-operation and Development, 13, 16

Paramagnetic resins, 73
Peptide nucleic acid, 171
Performance characteristics, 54–7
Performance control, 32–3
Performance parameters, 54
Pfu polymerase, 111
Pharmacogenomics, 213
Phenol, 69, 190
  as contaminant, 88
PicoGreen, 92
Pipettes, 124
Plan-Do-Check-Review, 18, 19
Plateau, 101–2
Plexor primers, 138–9
Polymerase chain reaction, 101–31
  amplification, 2
  amplification protocol, 103–5
  contamination control, 122–8
    decontamination methods, 124, 128
    dedicated equipment, 123–4
    physical laboratory separation, 123–4
    pipettes, 124
  deoxynucleotide triphosphates, 107
  DNA quantification, 93
  DNA template, 105–6
    concentration, 105–6
    integrity, 105
  enhancers, 113, 117–19
  features, 102
  gene-specific *see* Gene-specific PCR
  hot-start mechanisms, 110, 112
  inhibitors, 113–17
  magnesium chloride, 107
  multiplex *see* Multiplex PCR
  optimisation, 112–13
  post-PCR analysis, 125–7, 128
  primer design, 107–9
  quantitative *see* Quantitative real-time PCR
  reaction buffer, 106–7
  real-time *see* Real-time PCR
  target selection, 109
  thermal cycling, 119–22
    annealing, 119
    cycle number, 120
    denaturation, 119
    extension, 119–20
    ramp rate, 121–2
    temperature control, 120–1
    thermal cyclers, 120, 122
  thermal profile, 104
  thermostable DNA polymerases, 109–10, 111
  water quality, 107
Polyphenols, 8
Polyvinyl(poly)pyrrolidine, 72
Precision, 44–6
  intermediate, 46
  repeatability, 45
  reproducibility, 46
Primer extension pre-amplification, 180
  improvements in, 180
Primers, 101
  design, 107–9
    multiplex PCR, 170–1
    quantitative real-time PCR, 145–6
  optimisation, 149–50
  Plexor, 138–9
  positioning, 169–70
  Scorpion, 138
Principles of Good Laboratory Practice, 13, 16
Probes
  design, 145–6
  hybridisation, 136–8
  optimisation, 149–50
Proficiency testing, 5, 15, 28
Pronase, 70
Proteinase K, 70
Protein contamination, 87–8
Pulsed field gel electrophoresis, DNA requirements, 64

Quality assurance, 2, 3, 6, 12
Quality Control, 3, 6, 10–39
  contamination control, 33
  in-house control materials, 31–2
  microarrays, 226–9
    printing, 227–8
    spike-in controls, 229
    universal reference RNA, 228
  performance control, 32–3

reference materials, 31
RNA samples, 189–207
Quality Manual, 19–20, 21
Quality metrics, 234–5
Quality Procedures, 20
Quantification limits, 44
Quantitative real-time PCR, 103, 132–66
  assay design, 144–6
    probe and primer design, 145–6
    target sequence, 145
  contamination control, 152–3
  cycling conditions, 147–9
  data analysis, 154–6
    amplification efficiency, 156
    mathematics, 155
    normalisation, 155
    outliers, 156
  DNA requirement, 64
  experimental design, 153–4
  instrumentation, 141–4
  low target analyte levels, 157–61
    amplification cycles, 159
    data handling, 159–61
    replication level, 159
    sample handling, 158–9
  PCR master mix, 146–7
  primer and probe optimisation, 149–50
  product detection, 133–41
    5' nuclease assay, 134–5
    choice of fluorophores, 140–1
    hybridisation probes, 136–8
    melting curve analysis, 139–40
    molecular beacons, 135–6
    Plexor primer technology, 138–9
    Scorpion primers, 138
  standards and compatibility, 161–3
  target level, 150–2
  validation, 156–7
Quenchers, 137

Randomisation, 30
Reaction buffers, 106–7
Reagents, 25–6
  quality, 25
  stability/batch compatibility, 26
  storage conditions, 26

traceability, 26
Real-time PCR *see* Quantitative real-time PCR
Recording, 35–6
  automated analysis, 35–6
  documentation, 38
  electronic data, 35–6
Recovery, 47–8
Reference blanks
  fluorescence spectroscopy, 90
  ultraviolet spectroscopy, 86
Reference materials, 31
  certified, 6, 31
Reliable measurement, 4–6
Repeatability, 45, 57
Replicates, 30
Reporting, 4, 36
Reproducibility, 46
Restriction fragment length polymorphism, DNA requirements, 64
Reverse transcription PCR
  experimental design and data quality, 201–4
  RNA for *see* RNA
Review, 12
RNA
  as contaminant, 87
  extraction, 189–92
    additional purification, 190–1
    archival tissue samples, 191–2
    freezing, 189–90
    guanidinium isothiocyanate, 190
    phenol, 190
    sulfate, 190
  quality, 192–8
    inhibitors, 194–8
    spectrophotometric measurement, 194
  quantification, 199–201
    methods, 201
    significance of, 199–201
  universal reference, 228
RNA integrity number, 193–4
Robustness, 42–3, 54
  testing, 49
Ruggedness *see* Robustness

Subject Index

Safety, 22–3
Samples, 33–5
　archival, RNA extraction from, 191–2
　assessment, 4
　biological contamination, 7
　chain of custody, 33–4
　DNA extraction, 61
　freezing, 189–90
　uniqueness of, 42
Sample preparation, 34
　cell/nucleic adherence to matrix
　　material, 67
　contamination, 67
　DNA extraction, 66–7
　fluorescence spectroscopy, 89
　homogeneity of sample, 67
　surface area to lysis forces ratio, 67
　ultraviolet spectroscopy, 86
Sample storage, 34–5
　DNA extraction, 66
　environment, 66
　temperature, 66
Sarkosyl, 69
Scorpion primers, 138
Selectivity, 49–50
Sensitivity, 50–1
Short tandem repeat analysis, DNA
　requirements, 64
Short tandem repeats, 8
Silica particles, 73
Sodium dodecyl sulfate, 69
Solid phase capture, 126
Spatial separation, 23–4
Spike-in controls, 229
Staff, 26–7
　culture and competence, 26
　qualifications, 5
　training and development, 26–7
Standard Operating Procedures, 20–2
Stoffel fragment, 111
Sulfate solutions, 190
SYBR dyes, 92

T7-based linear DNA amplification,
　181–2

TaqMan assay, 134–5
Technical approach, 4
Testing and calibration standards, 15–16
Thermal cyclers, 120, 122
Thermus aquaticus (*Taq*) polymerase,
　109–10, 111
Topo Taq, 111
Toxicogenomics, 213
Traceability, 10
　metrological, 10, 28
　reagents, 26
Trackability, 10
Training, 26–7
Triton-X detergents, 69
Tween detergents, 69

Ultrafiltration, 73
Ultraviolet light, 128
Ultraviolet spectroscopy, 84–8
　calibration of spectrophotometer, 85
　cuvettes, 85–6
　extinction coefficient, 84–5
　light source, 86
　presence of contaminants, 87–8
　reference blank, 86
　sample dilution, 86
　sample preparation, 86
United Kingdom Accreditation
　Service, 15
Uracil-N-glycosylase, 124

Valid Analytical Measurement, 4–5
Validation, 2, 4, 29, 40–58
　definition, 40
　method performance parameters,
　　43–53
　　accuracy, 48
　　bias, 46–7
　　detection limit (sensitivity), 50–1
　　measurement uncertainty, 52–3
　　precision, 44–6
　　recovery, 47–8
　　ruggedness (robustness) testing, 49
　　selectivity, 49–50
　　working range and linearity, 51–2

necessity for, 41–3
  criticality of data, 42
  expected level of utilisation of
    technique, 43
  robustness of method, 42–3
  uniqueness of sample, 42
practical application, 53–7
  analytical requirement, 53–4
  critical parameters, 54
  documentation, 57
  draft protocol, 54
  fitness of method, 55–7
  limitations of method, 57
  performance characteristics, 54–5
  performance parameters, 54
process planning, 43
quantitative real-time PCR, 156–7
VENT polymerase, 111

Verification, 42
  see also Validation

Whole genome amplification, 179–84
  applications and characteristics, 182–4
  degenerate oligonucleotide primed
    PCR, 179–80
  improved primer extension pre-
    amplification, 180
  ligation-mediated PCR, 181
  multiple displacement amplification,
    180–1
  primer extension pre-amplification, 180
  T7-based linear DNA
    amplification, 181–2
Working range, 51–2

Zymolase, 70